MATHEMATICAL STATISTICS AND APPLICATIONS

MATHEMATICAL STATISTICS AND APPLICATIONS

Proceedings of the 4th Pannonian Symposium on Mathematical Statistics, Bad Tatzmannsdorf, Austria, 4–10 September, 1983

Volume B

Edited by

WILFRIED GROSSMANN

Institute of Statistics and Computer Science, University of Vienna, Austria

GEORG Ch. PFLUG

Institute of Mathematics, University of Gießen, FRG

ISTVÁN VINCZE

Mathematical Institute, Hungarian Academy of Sciences, Budapest, Hungary

WOLFGANG WERTZ

Institute of Statistics, Technical University of Vienna, Austria

D. REIDEL PUBLISHING COMPANY

A MEMBER OF THE KLUWER ACADEMIC PUBLISHERS GROUP

DORDRECHT / BOSTON / LANCASTER / TOKYO

Library of Congress Cataloging-in-Publication Data

Pannonian Symposium on Mathematical Statistics (4th:
1983: Bad Tatzmannsdorf, Austria)
Mathematical statistics and applications.

(Proceedings of the 4th Pannonian Symposium on Mathematical Statistics, Bad Tatzmannsdorf, Austria, 4–10 September 1983; v. B)
Bibliography: p.
Includes index.
1. Mathematical statistics—Congresses. I. Grossmann, Wilfried, 1948– . II. Series: Pannonian Symposium on Mathematical Statistics (4th: 1983: Bad Tatzmannsdorf, Austria). Proceedings of the 4th Pannonian Symposium on Mathematical Statistics, Bad Tatzmannsdorf, Austria, 4–10 September 1983; v. B.
QA276.A1P36 1983 vol. B 519.5 s [519.5] 85-14464

ISBN-13:978-94-010-8901-2 e-ISBN-13:978-94-009-5438-0
DOI:10.1007/978-94-009-5438-0

Distributors for the U.S.A. and Canada
Kluwer Academic Publishers,
190 Old Derby Street, Hingham, MA 02043, U.S.A.

Distributors for Albania, Bulgaria, Chinese People's Republic,
Cuba, Czechoslovakia, German Democratic Republic, Hungary,
Korean People's Republic, Mongolia, Poland, Romania,
the U.S.S.R., Vietnam and Yugoslavia
Kultura Hungarian Foreign Trading Company
P.O.B. 149
H-1389 Budapest 62, Hungary

Distributors for all remaining countries
Kluwer Academic Publishers Group,
P.O.Box 322, 3300 AH Dordrecht, Holland

Joint edition published by D. Reidel Publishing Company
Dordrecht, Holland
and
Akadémiai Kiadó, Budapest, Hungary

Softcover reprint of the hardcover 1st edition 1985

CONTENTS

PREFACE

The Fourth Pannonian Symposium on Mathematical Statistics was held in Bad Tatzmannsdorf, Austria, 4-10 September, 1983. The first two Symposia were held there in 1979 and 1981, whereas the Third Symposium was staged in Visegrád, Hungary in 1982. The proceedings volumes of these conferences, published by Springer, D. Reidel, and D. Reidel & Akadémiai Kiadó, respectively give information about the objectives of the Pannonian Symposia and the topics covered.

About 130 participants from 17 countries took part in this Fourth Symposium, and 92 lectures were presented. This volume contains 21 reviewed contributions which cover various aspects of the application of mathematical statistics. A second group of papers dealing with problems of probability theory and decision theory is published in a separate volume entitled "Probability and Statistical Decision Theory".

Roughly speaking, the papers can be grouped into three main categories. The first group is the application of probability theory. A special type of application is shown in the invited paper of P. Erdös, namely probabilistic methods in number theory. Further models of applied probability covered by the papers are game theory, urn models, best choice models and random graphs. The second group could be best characterized by the term mathematical statistics for models of real data. Such models are linear models, regression, discrimination, time series, analysis of censored data, goodness of fit approximation of processes. The papers show the increasing importance of new theoretical results (i.e. in nonparametric regression, robustness or limit theorems) to applied problems. The last group of papers shows various aspects of computational statistics and the application of Monte Carlo methods for numerical problems such as integration or optimization.

We wish to express our thanks to many persons who gave us indispensable aid in the preparation of this volume: J. Anděl, O. Bunke, I. Csiszár, L. Devroye, B. Gyires, J. Hurt, M. Hušková, T. Nemetz, E. Neuwirth, H. Niederreiter, D. Plachky, B. Pötscher, F. Pukelsheim, P. Révész, E. Ronchetti, H. Stetter, W. Stute, K. Urbanik, and R. Zieliński, for their help in refereeing the papers and in other editorial matters, the secretarial staff of Professor I. Kátai at the University of Budapest, who did the laborious typing of the manuscript under the supervision of Mrs. Z. Andrásné Králik and, last but not least, Akadémiai Kiadó and D. Reidel Publishers for their good cooperation. (After the refereeing process and retyping, all papers were returned to the authors for correction.)

The organization of the Symposium was made possible by the help of many individuals and institutions. The organizers gratefully acknowledge the generous support given by the State Government of Burgenland, the Federal Ministry of Science and Research, the Austrian Statistical Society, the Control Data Co., Hypobank of Burgenland Co., the Volksbank Oberwart Co., the Raiffeisenbank Oberwart Co., the Kurbad Tatzmannsdorf Co. and the Local Authority of Bad Tatzmannsdorf; special thanks go to Th. Kery and Dr. R. Grohotolsky (Head and Vice-Head of Burgenland, respectively), Dr. J. Karall (Member of the State Government of Burgenland), DDr. Schranz (Member of State Parliament of Burgenland) and Mag. R. Luipersbeck (Director of the Kurbad Tatzmannsdorf AG) for their help in many respects. Finally, we express our thanks to the ladies for their splendid work in the preparation and local organization of the meeting; in particular to Ms. Ingrid Danzinger who patiently undertook most of the typing for the organization.

Wolfgang WERTZ

Proc. of the 4th Pannonian Symp. on Math. Stat., Bad Tatzmannsdorf, Austria 1983
W. Grossmann, G. Pflug, I. Vincze, W. Wertz, eds

SOME APPLICATIONS OF PROBABILITY METHODS TO NUMBER THEORY

P. ERDŐS

Budapest

I wrote many papers on these subjects. In this note I will mainly state some unsolved problems which have perhaps been somewhat neglected and which perhaps cannot be solved completely by probability methods, but where slightly weaker results can be obtained by these methods. I will mainly discuss my own problems and those of my coworkers not because I consider these more important but because I hope I know more about them than the reader. Also since my memory is still good I will occasionally add some historical remarks.

First a few words about the literature, this list of references will be very incomplete.

P.D.T.A. Elliott, Probabilistic number theory, Springer Verlag, Grundlehren der Math. Wissenschaften, Vol. 239 and 240 (1980). This comprehensive handbook deals with additive and multiplicative arithmetical functions, it contains also many references to the earlier literature and many unsolved problems and interesting historical remarks.

H. Halberstam and K. F. Roth, Sequences, Springer Verlag 1982. The third chapter of this excellent book contains applications of probability methods to additive number theory.

The first survey paper on probability methods in number theory is M. Kac, Probability methods in some problems of analysis and number theory, Bull. Amer. Math. Soc. 55 (1949), 641-665. I feel that this interesting paper deserves careful study even now.

I will give references only if they are not contained in these books.

Before I start my subject I just remark that probabilistic ideas are often useful in making plausible conjectures which cannot be attacked by our methods which are at our disposal at present. The best known such conjecture is due to Cramer: Let $p_1<p_2<\ldots$ be the sequence of consecutive primes. Then

$$(1) \qquad \overline{\lim} \frac{p_{n+1}-p_n}{(\log n)^2} = 1.$$

The Riemann hypothesis would only imply $p_{n+1}-p_n<p_n^{\frac{1}{2}+\varepsilon}$ and the currently known best inequality from below due to Rankin states that there is a $c>0$ so that for infinitely many n

$$(2) \quad p_{n+1}-p_n > \frac{c \log n \ \operatorname{loglog} n \ \operatorname{loglogloglog} n}{(\operatorname{logloglog} n)^2}$$

Rankin obtained (2) 45 years ago, no progress has been made since then except that Schönhage and Rankin himself improved the value of c. This prompted me to offer a reward of 10000 dollars for a proof that (2) holds for every c and infinitely many n. I am so sure that (2) holds that for a disproof I offer 25000 dollars. The only reason that I do not offer 10^6 dollars that is the very unlikely event that I am wrong and (2) does not hold for every c I could not pay my debt.

Let me state two less well known conjectures of myself: Denote by $P(m)$ the greatest prime factor of m. Is it true that for every $n>n_0(\varepsilon)$ $P(n(n+1)) > (\log n)^{2-\varepsilon}$, but for infinitely many n $P(n(n+1)) < (\log n)^{2+\varepsilon}$. Also is it true that every $n > n_0(\varepsilon)$ can be written in the form $n=a+b$ $P(a\cdot b) < (\log n)^{2+\varepsilon}$? It is easy to see that the result fails if we replace $(\log n)^{2+\varepsilon}$ by $(\log n)^{2-\varepsilon}$. Very much weaker positive results have been proved by analytic methods by Balog and Sárközy. Balog and Sárközy will publish several papers on this and related subjects.

A. Balog and A. Sárközy, On sums of integers having small prime factors, I-II, Studia Sci. Math. Hung., and On sums of sequences of integers, I-III, Acta Arithmetica and Acta Math. Acad. Sci. Hung., to appear.

H. Cramer, On the order of magnitude of the difference between consecutive prime numbers, Acta Arith. 2 (1936), 23-46.

R. A. Rankin, The difference between consecutive prime numbers, J. London Math. Soc. 13 (1938), 242-247.

First a few words about additive and multiplicative number theoretic functions, this chapter will be very short in view of the book of Elliott. I proved many years ago that the density of integers n for which $\varphi(n) > \varphi(n+1)$ is $\frac{1}{2}$ and the same result holds for $\sigma(n)$ and for $d(n)$ the number of divisors of n. This later result was a conjecture of Chowla. I could never prove that the density of integers with

$P(n) > P(n+1)$ is $\frac{1}{2}$ and this conjecture is probably unattackable by methods at our disposal. Very much weaker results have been proved by Pomerance and myself.

On the other hand easy independence arguments will give that the density of integers for which $\varphi(n) > \varphi(n+1)$ and $d(n) > d(n+1)$ is $\frac{1}{4}$ since the two inequalities are asymptotically independent. Put $f(n) = \sum_{p|n} \frac{1}{\log\log p}$. With very little more trouble I can prove that the three inequalities $d(n) > d(n+1)$, $f(n) > f(n+1)$ and $\varphi(n) > \varphi(n+1)$ are asymptotically independent. On the other hand $\varphi(n) > \varphi(n+1)$ and $\sigma(n) < \sigma(n+1)$ are strongly correlated. The density of integers which satisfy both inequalities is strictly between $\frac{1}{4}$ and $\frac{1}{2}$. Finally the density of integers with $d(n) > d(n+1)$, $\omega(n) > \omega(n+1)$ is $\frac{1}{2}$ ($\omega(n)$ denotes the number of distinct prime factors of n). I am not sure if these results are in the literuture but anyone familiar with the methods of probabilistic number theory can easily supply the proofs. I just want to state one of my old problems which does not seem quite hopeless but which so far resisted all attacks. Let $f(n)$ be an additive function and assume that for a pair a, b of real numbers the density of integers n for which $a<f(n)<b$ exists and is positive. Does it then follow that $f(n)$ has a limiting distribution? It is not difficult to prove that our condition implies that the two series

$$\sum_{|f(p)|>1} \frac{1}{p}, \qquad \sum_{|f(p)|<1} \frac{f(p)^2}{p}$$

both converge, but I could not prove that

$$\sum_{|f(p)|<1} \frac{f(p)}{p}$$

also converges. If this would also be proved then by the theorem of Wintner and my conjecture would be settled. Elliott gave a purely probabilistic formulation of my conjecture (see Vol. 2, p. 331 of Elliott's book).

An old conjecture of mine on multiplicative functions stated

Let $f(n)$ be a multiplicative function which only takes the values ± 1. I conjectured that $f(n)$ has a mean value, i.e. that

$$(3) \qquad \lim \frac{1}{x} \sum_{n=1}^{x} f(n)$$

exists, and is 0 if and only if $\sum_{f(p)=-1} \frac{1}{p} = \infty$. (3) was first proved by Wirsing and later in a more general form by Halász. Tchudakoff and I conjectured (independently) that for such a multiplicative function

$$(4) \qquad \overline{\lim} \left| \sum_{n=1}^{x} f(n) \right| = \infty.$$

In fact I have a more general conjecture: Let $f(n) = \pm 1$ be any number theoretic function (not necessarily multiplicative) then to every c there is a d and m so that

$$(5) \qquad \left| \sum_{k=1}^{m} f(kd) \right| > c.$$

(5) was one of my first conjectures and is now more than 50 years old. I offer 500 dollars for a proof or disproof of (5).

Choose $f(p) = +1$ or $f(p) = -1$ with probability $\frac{1}{2}$ and for $n = \Pi p_i^{\alpha_i}$ put $f(n) = \prod_{p|n} f(p)^{\alpha_i}$. Wintner asked what can be said about $\sum_{n=1}^{x} f(n)$ for almost all choices of $f(p) = \pm 1$. He proved that for almost all functions $f(n)$ we have

$$(6) \qquad \sum_{n=1}^{x} f(n) < x^{\frac{1}{2}+\varepsilon}.$$

I improved (6) and showed that for almost all functions $f(n)$

$$c_1 \left(\frac{x}{\log x}\right)^{1/2} < \sum_{n=1}^{x} f(n) < x^{1/2} (\log x)^{c_2}$$

and conjectured that for almost all functions

$$(7) \qquad \overline{\lim} \frac{1}{x^{1/2}} \sum_{n=1}^{x} f(n) = \infty$$

but

$$(8) \qquad \lim \frac{\sum_{n=1}^{x} f(n)}{x^{1/2} (\log x)^{\varepsilon}} = 0$$

I could not even guess the analog of the law of the iterated logarithm. Halász proved (8), but only proved a slightly weaker result than (7). As far as I know nobody has a plausible guess for the true order of magnitude of $\left|\sum_{n=1}^{x} f(n)\right|$.

Many further interesting questions could be asked e.g.: What can be said about the number of 0-s of the partial sums $\sum_{n<x} f(n)$ for the random multiplicative function? By analogy with the Rademacher functions one would expect that the number of zeros is between $x^{\frac{1}{2}-\varepsilon}$ and $x^{\frac{1}{2}+\varepsilon}$,

of course more precise results would be very desirable.

G. Halász, On random multiplication functions, Publ. Math. D'Orsay 1983, Journées Arith. Coll. H. Delange 79-96.

P. Erdös and Carl Pomerance, On the largest prime factors of n and n+1, Aequationes Math. 17 (1978), 311-321.

Now I discuss some problems and results on additive number theory. These problems were first stated by Sidon more than 50 years ago, he was led to these problems by his study of lacunary trigonometric series. Let $a_1 < a_2 < \ldots$ be an infinite sequence of integers, denote by $f(n)$ the number of solutions of $n=a_i+a_j$. Sidon asked me in 1932 when we first met whether there is a sequence $A = \{a_1 < a_2 < \ldots\}$ for which $f(n) > 0$ for all n but for which $f(n)/n^{\varepsilon} \to 0$ for every $\varepsilon>0$ i.e. A is a basis of order 2 but $f(n)$ is small. I at first thought that the problem will not be hard and that it will be easy to construct such an A. I never succeeded in constructing such an A but about 20 years later I proved by probabilistic methods that there is a sequence A for which

(9) $$c_1 \log n < f(n) < c_2 \log n$$

holds for every n. An outstanding problem here is whether there is a sequence A for which

(10) $$f(n) = (1 + o(1))\log n$$

I expect that such a sequence A does not exist. As a first step one should prove that for every sequence A

(11) $$\overline{\lim}\ |f(n) - \log n| = \infty$$

(11) will perhaps not be hard to prove. I offer 500 dollars for a proof or disproof of (10). If there is no A satisfying (10) then one could ask: Put

$$\overline{\lim}\ f(n)/\log n = C_1, \ \underline{\lim}\ f(n)/\log n = C_2$$

Is there an $\varepsilon>0$ so that for every A, $C_1/C_2 > 1+\varepsilon$?

This question just occured to me while I write these lines and I hope it will not turn out to be trivial. Let g(n) be a monotonic function which tends to infinity arbitrarily slowly. It is easy to prove by the probability method that there is a sequence A for which

$$f(n)/g(n)\ \log n \to 1.$$

An old conjecture of Turán and myself states that if

(12) $f(n) > 0$ for all n then $\overline{\lim}\ f(n) = \infty$

I offer 500 dollars for a proof or disproof of (12).

Perhaps $f(n) > 0$ for all $n > n_0$ already implies that there is an absolute constant c so that $f(n) > c \log n$. If this conjecture is true one can again ask: Is it true that in fact $c > c_0$?

Another possible strengthening of my conjecture with Turán would be: Assure $a_k < c\, k^2$. Is it then true that $\overline{\lim}\, f(n) = \infty$ and perhaps even $f(n) > c \log n$ for infinitely many n?

Sidon calls an infinite sequence A a B_2 sequence if the integers $a_i + a_j$ are all distinct. When we first met Sidon asked for a B_2 sequence for which a_n increases as slowly as possible. The greedy algorithm easily gives that there is a B_2 sequence for which $a_k < c\, k^3$ and we both conjectured that in fact there is a B_2 sequence for which $a_k < k^{2+\varepsilon}$. Rényi and I proved by probabilistic methods that there is a sequence A with $a_k < k^{2+\varepsilon}$ and $f(n) < C_\varepsilon$ for all n. For nearly 50 years we could not prove that there is a B_2 sequence for which $a_k = o(k^3)$. A few years ago Ajtai, Komlós and Szemerédi proved by an ingenious combination of combinatorial and probabilistic methods that there is a B_2 sequence for which

$$(13) \qquad a_k < c\, k^3 / \log k.$$

(13) at the moment seems to be the natural boundary of their method.

Let me state an interesting problem in connection with the greedy algorithm. We construct a B_2 sequence as follows:

Assume $a_1<a_2<\dots<a_{k-1}$ has already been constructed. Then a_k is the smallest integer for which $\{a_1,\dots,a_k\}$ is a B_2 sequence. It is easy to see that $a_k<c\,k^3$ and Chowla and Mian carried out extensive calculations on the basis of which they suggest that $a_k \sim k^{2+c}$. It is not even known if

$$(14)\quad a_k/k^2 \to \infty \qquad \text{and} \qquad a_k/k^3 \to 0.$$

I am not an expert on algorithms but find (14) a fascinating conjecture and offer 250 dollars for a proof or disproof.

I proved that if A is a B_2 sequence then

$$(15)\qquad \overline{\lim}\ \frac{a_k}{k^2 \log k} > c > 0$$

for a certain $c>0$. Is (15) best possible? I have no information and offer 500 dollars for a proof or disproof. It is best possible in the following much weaker sense. I proved (15) by showing that if $a_k < c_1 k^2 \log k$ for all $k > k_o$, c_1 sufficiently small, then the number of solutions $g(x)$ of $0 < a_j - a_i < x$ satisfies $g(x) > x$. Thus our sequence cannot be a B_2 sequence. It is not hard to show that (15) is best possible in this case i.e. if $a_k = C\,k^2 \log k$, C sufficiently large, then $g(x)<x$. Krickeberg and I proved that there is a B_2 sequence for which

$$(16)\qquad \underline{\lim}\ a_k/k^2 < C$$

for some C. The best result in (16) is due to Krickeberg:

$C > \frac{1}{\sqrt{2}}$. The "Truth" is probably C=1. Perhaps if the $\underline{\lim}$ in (16) is finite then the $\overline{\lim}$ in (15) is infinite. As far as I know this question has not yet been investigated.

Straus and I conjectured that if A is any infinite sequence then there always is a sequence B of density 0 so that every integer is of the form A+B (i.e. of the form a_i+b_j). In fact Lorentz proved the following stronger result: Denote $A(x) = \sum_{a_i<x} 1$. Then there always is a sequence B for which A+B is I e.i. the sequence of all integers and

$$(17) \qquad B(x) < c \sum_{n=1}^{x} \frac{\log A(n)}{A(n)}$$

The proof of (17) is nonconstructive and Lorentz remarks that it often is close to being best possible. Lorentz in particular deduces from (17) that if A is the sequence of primes then $B(x) < c(\log x)^3$. I proved by probabilistic methods that in fact $B(x) < c(\log x)^2$ is also possible. I think it would be very interesting to decide if for every additive complement of the primes we in fact have

$$(18) \qquad \overline{\lim} \frac{B(x)}{\log x} = \infty$$

and if there is an additive complement for which

$$(19) \qquad \lim B(x)/(\log x)^2 = 0$$

I could not even prove that there is no additive complement of the primes with

(20) $\lim B(x)/\log x = 1$

I am certain that no such sequence exists.

I offer 100 dollars for a proof that there is no additive complement satisfying (20) and 1000 dollars for a proof that such a sequence exists. Probability methods probably will not help and at present no methods seem to be available.

Lorentz observed that (17) implies that if A has positive density then there is an additive complement B for which $B(x) \leq c(\log x)^2$. I showed that this is best possible in general. Let Q be the sequence of squarefree numbers. It is easy to see from the chinese remainder theorem that if F is an additive complement of Q then for a certain c>0

(21) $$B(x) > \frac{c \log x}{\log\log x}.$$

It is not difficult to prove by probabilistic methods that there is an additive complement B for which $B(x) < c \log x$. It would be of some interest to determine the exact order of B(x) in this case.

It is easy to show by the probability method that almost all sequences A have an additive complement B for which $B(x) = (1+o(1))\frac{\log x}{\log 2}$. The measure in the space of sequences is the Lebesgue measure i.e. we make correspond to the sequence A the real number $\sum_{i=1}^{\infty} \frac{1}{2^{a_i}}$

Just a few words about problems and results of Nathanson and myself. We proved by probabilistic methods that there is a sequence A for which

$$c_1 \log n < f(n) < c_2 \log n$$

and if A(n) denotes the set of integers a+b=n then $|A(n) \cap A(m)| < 4$ for every n and m, we could not decide if 4 can be replaced by 3 and perhaps even by 2. More generally the probabilistic method gives that there is a basis A of order k for which $A(x) < cn^{1/k}(\log n)^{1/k}$ and for which if A (n) denotes the set of a 's which occur in the set of solutions

$$n = \sum_{i=1}^{s} \varepsilon_i a_i, \sum_{i=1}^{s} \varepsilon_i < k, 0 < \varepsilon_i < k, \varepsilon_i \text{ integers}$$

then

$$|M_k(n) \cap M_k(m)| < C_k$$

The exact value of C_k is unknown, as just stated $2<C_2<4$.

We further asked: Is there a basis of order 2 of density 0 for which the equation $a_i+a_j = a_r$ has only a finite number of solutions? The odd numbers show that the condition "density 0" cannot entirely be dropped. More generally we asked: Is there for every k a basis of order k for which all the sums $\sum_{i=1}^{r} \varepsilon_i a_i$, $\varepsilon_i=0$ or 1, are all distinct (except for a finite number of cases) as long as $r \leq k-1$?

Finally I want to mention some early work of Atkin. Littlewood posed the following problem: Does there exist a sequence of integers $a_1 < a_2 < \ldots$ for which $|a_k - k^2| < (\log k)^c$ and for which the sequence $a_i + a_j$ has positive density? Atkin proved sometime before 1950 that the answer is positive and that in fact we can choose $c=1$. Atkin uses semi-probabilistic methods. He told me of his results in 1949 during my visit to England. I unfortunately forgot completely about this and made therefore no reference ot Atkin's unpublished results in my papers on the probability method around 1954-55. Later I remembered my conversations with Atkin and we referred to his work in my papers with Rényi. It is of course impossible to tell if my conversations with Atkin had influenced my later papers. I believe I can prove that the best possible value of c in Littlewood's problem is log 2 but I never worked out all details in full.

A.O.L. Atkin, On pseudo-squares. Proc. London Math. Soc., Third series 14 (1965), 22-27.

M. Ajtai, J. Komlós and E. Szemeredi, On dense infinite Sidon sequences, European J. Comb. 2 (1981), 1-11.

See also

M. Ajtai, P. Erdös, T. Komlós and E. Szemerédi, On Turán's theorem for sparse graphs, Combinatorica 1 (1981), 313-317.

J. Komlós, J. Pintz and E. Szemerédi, A lower bound for Heilbronn's problem, J. London Math. Soc. 25 (1982), 13-24.

To end this paper I discuss some problems which I have somewhat neglected. First of all an old problem of mine stated Divide the integers $1 \le k \le 2n$ into two disjoint subsets $a_1 < a_2 < \ldots < a_n$; $b_1 < b_2 < \ldots < b_n$. Put

$$M_k = \sum_{a_i - b_j = k} 1, \quad M = M(n) = \min \max_k M_k$$

where the minimum is to be taken over all possible divisions of $1 \le k \le 2n$ into two sets of size n A and B. I first thought that $M = \frac{n}{2}$ but showed by the probability method that $M < \frac{4}{9} n$. Independently and about simultaneously Selfridge, Motzkin and Ralston showed by aid of an early electronic computer SWAC that for $n=15$ $M=6$ and they observed that this implies that for infinitely many n $M \le 0.4\, n$. I further showed $M \ge 0.25\, n$ and Scherk improved this to $M > (1 - \frac{1}{\sqrt{2}})n$. Finally L. Moser showed that

$$M > (4-15^{1/2})^{1/2} (n-1) > 0.3970\ (n-1).$$

It would perhaps be worthwhile to get the best possible value of M or at least to determine the smallest c for which $M \ge cn + o(n)$.

L. Moser: On the overlap problem of Erdös, Acta Arith. 5 (1959), 117–119.

An old result of Tchebisheff states that the probability that n and m are relatively prime is $\frac{6}{\pi^2}$. One can expect that this will remain true in general for a large class of number theoretic functions $g(n)$. E. g. R. R. Hall proved this if $g(n) = \omega(n)$, the number of distinct prime factors of n. Probably it will be true for every $[n\alpha]$, $0<\alpha<\infty$, α not an integer and this will probably not be very difficult to prove (if it is not already in the literature). On the other hand it is probably hopeless to prove that for all $\alpha>1$, α not an integer, the density of integers n for which $(n,[\alpha^n])=1$ is $\frac{6}{\pi^2}$. I do not see how to show this for almost all α, but it is doubtful if probability methods will be of any help here.

P. Erdős and G. G. Lorentz, On the probability that n and $g(n)$ are relatively prime, Acta Arith. 5 (1959), 35-44.

R. R. Hall, On the probability that n and $f(n)$ are relatively prime, Acta Arith. 17 (1970), 169-183.

Rényi and I proved the following theorem. Let G be an additively written Abelian group of order n . Let us choose k elements of G at random, consider all the 2^k sums

$$(22) \qquad \sum_{i=1}^{k} \varepsilon_i a_i, \ \varepsilon_i = 0 \text{ or } 1.$$

Rényi and I proved that if

$$(23) \qquad k > \frac{\log n + \log\log n + \omega_n}{\log 2}$$

where $\omega_n \to \infty$ arbitrarily slowly then almost surely all elements of G can be represented in the form (22). We further showed that if

$$(24) \qquad k > \frac{(2 + o(1))\log n}{\log 2}$$

then almost surely all elements of G have

$$(25) \qquad (1 + o(1)) \frac{2^k}{n}$$

representations in the form (22). K. Bognár improved (24) and later R. R. Hall and I proved that in (24) $2+o(1)$ can be replaced by $1+o(1)$.

It is curious that these results hold for every Abelian group of order n, in other words the results are independent of the structure of G. Almost certainly (23) is best possible if we insist that the result should hold for all Abelian groups of order n independent of the structure, but as far as I know this question has not yet been completely cleared up. Also the best possible error term in the result of R. R. Hall and myself is very far from being known.

These results have a surprising number theoretic application. Denote by $f(x,k,\ell)$ the number of integers $n<x$ which have a divisor

$$d \equiv \ell \pmod k, \quad (k,\ell) = 1,$$

and $F(x;k)$ denotes the number of integers $n<x$ which have a divisor in every residue class $\ell \pmod k$ $(\ell,k = 1$. Clearly $F(x;k) \leq f(x;k,\ell)$. I proved that for $k<2^{(1-\varepsilon)\log\log x}$

(26) $$F(x;k) = x + o(x)$$

uniformly in k.

The following problem is perhaps of some interest: Let $k(n) = k$ be the smallest integer for which there is an $\ell, (\ell,k) = 1$ for which n has no divisor $d \equiv \ell \pmod{k}$. (26) implies that the normal order of $k(n)$ is $(\log n)^{\log 2}$, but one could try to obtain sharper results and one could try to estimate $\max_{n<x} k(n)$. I hope to return to these and related questions in the future if there is a future for me.

P. ERDÖS, On the distribution of divisors of integers in the residue classes mod d.

P. ERDŐS and A. RÉNYI, Probalilistic methods in group theory, J. Analyse Math. 14 (1965), 127-138.

K. BOGNÁR, On a problem of statistical group theory, Studia Sci. Math. Hungar. 5 (1970), 29-36.

P. ERDŐS and R. R. HALL, Probabilistic methods in group theory, Houston J. of Math. 2 (1976), 173-180,

and Some results in probabilistic group theory, Comment. Math. Helv. 53 (1978), 448-457. This paper contains many further references.

Pál Erdős
Mathematical Institute of the
Hungarian Academy of Science
Reáltanoda utca 13-15.
H-1364 BUDAPEST

Proc. of the 4th Pannonian Symp.
on Math. Stat., Bad Tatzmannsdorf, Austria 1983
W. Grossmann, G. Pflug, I. Vincze, W. Wertz, eds

ON AN INFORMATIONTHEORETICAL APPROACH FOR A CERTAIN CLASS OF GAMES

JOHANNES ADLER

Wien

SUMMARY:

We consider the following game where each player bets on the outcome of certain mutually exclusive and exhaustive events $x_1,\dots,x_m$ with probabilities $p_1,\dots,p_m$ respectiveley. Given that $\xi=x_i$ occured, the bank's payoff to each player is proportional to the stake deposited on x_i by that player Usually the bank just pays 100q percent, $q\leq 1$, of its total income out. Nevertheless it is profitable to participate such a game even in cases where q is small, if the other players behave sufficiently irrational. We investigate under what circumstances the game is profitable and give the optimal strategy, where the criterion for the choice of that strategy is rather based on risk aversion.

1. THE GAME

Most of the usual casino games have predetermined quotas. If a player bets on the outcome of a certain event x_i having probability p_i to occur he wins back a predetermined multiple $\frac{1}{q_i}$ of the stake he has deposited on x_i. Betting on x_i is considered fair if $p_i=q_i$ and unfair if $p_i<q_i$. A typical example of this kind is the roulette. In this game the income of the casino depends on chance, that is on the actual realization of the gamble. The gambler's fortune doesn't depend on the strategy of the other players; each player plays solely against the bank.

A different type of gamble is described by the following rules: the players bet on the occurrence of elements of a set $X = \{x_1,\ldots,x_m\}$, having probabilities $P = \{p_1,\ldots,p_m\}$, $\sum_{i=1}^{m} p_i=1$. The bank takes from every player $100(1-q)$ percent of his stake and divides the remaining $100q$ percent among the winners. Consider a play of the game: if the outcome is x_i and N_i units were deposited on x_i, whereas $N = \sum_{k=1}^{m} N_k$ units were deposited in that play, then for each unit deposited on x_i the player gets $q \frac{N}{N_i}$ units back. In this case the player wins $\frac{1}{q_i} = q \frac{N}{N_i}$ times the stake deposited on x_i.

Unlike the previous example, the quotas $\frac{1}{q_i}$ depend on the way the other players are betting: denote by $100q_i{}^* = 100 \frac{N_i}{N}$ the percentage of the total stake that has been deposited on x_i to obtain $q_i = \frac{q_i{}^*}{q}$.

We thus have another distribution $Q = \{q_1,\ldots,q_m\}$ on X, with $\sum_{i=1}^{m} q_i = \frac{1}{q} \geq 1$.

If q=1 the bank keeps no money and we thus have a zero-sum game; if q<1 and we consider the bank not as a player we have not a zero-sum game. Considering a specific player we note that he is actually not playing against the bank but against all other players.

It is now our aim to give for this player an optimal strategy, which is given by a portfolio $B = \{b_0,b_1,\ldots,b_m\}$, $b_i \geq 0$, $\sum_{i=0}^{m} b_i=1$, with the interpretation that b_i is the proportion of capital allocated to outcome x_i, $i=1,\ldots,m$ and b_0 is the proportion of capital the player doesn't stake.

The game is thus given by the set X, a number q, $0<q\leq 1$ and two probability distributions on $X: P$, $Q^* = \{q_1^*, \ldots, q_m^*\}$, $q_i^* = qq_i$. Denote by ξ_n the outcome of the n-th play of the game and by S_0 the initial capital of the player. The fortune after the n+1-st play is given by

$$S_{n+1} = \frac{b_i}{q_i} S_n + b_0 S_n \qquad \text{if} \quad \xi_{n+1} = x_i$$

$$= (\frac{b_i}{q_i} + b_0)\, S_n = r_i\, S_n.$$

Thus $S_{n+1} = r_{\xi_{n+1}} S_n = r_{\xi_{n+1}} \ldots r_{\xi_1} S_0$ holds.

Since the different plays of the game are supposed to be independently and identically distributed it is clear that an optimal strategy should neither depend on the number of the play nor on the outcomes of the previous ones, see Breiman (1961). Since we are interested how many percent of the initial capital are left after the n-th play we set $S_0=1$.

2. CHOICE OF THE OPTIMAL STRATEGY

On what criterion should we choose the optimal portfolio B?

1. Utility function.

   Suppose the gambler has a logarithmic utility for money, which means that he behaves risk averse since log is a concave function. Moreover he avoids going bankrupt under all circumstances since $\log 0 = -\infty$. The criterion for the choice of B is maximizing the expected utility

$$E \log S_n = n\, E \log r_{\xi_1} = n \sum_{i=0}^{m} p_i \log \left(\frac{b_i}{q_i} + b_o\right).$$

By Jensen's inequality we have

$$E\, S_n \geq \exp(E \log S_n) = \exp(n \Sigma p_i \log(\frac{b_i}{q_i} + b_o))$$

and therefore $E\, S_n > 1$ if $\Sigma p_i \log(\frac{b_i}{q_i} + b_o) > 0$. $E\, S_n$ increases at least geometrically fast in that case.

2. Asymptotic average interest rate.

$r_{\xi_n} - 1$ is the interest rate of the player for the n-th party of the game, $\sqrt[n]{S_n}-1$ is the average interest rate for one party during the first n partys of the game and

$$\lim_{n\to\infty} \sqrt[n]{S_n}-1 = \lim_{n\to\infty} \exp(\frac{1}{n} \log S_n) - 1 = \exp(E \log r_{\xi_1})-1$$

is the asymptotic average interest rate, whereas the last equality follows by the strong law of large numbers. The asymptotic interest rate depends on B and maximizing the asymptotic interest rate is equivalent to maximizing

$$E \log r_{\xi_1} = \sum_{i=1} p_i \log \left(\frac{b_i}{q_i} + b_o\right) = f(B)$$

over the m-dimensional simples Σ_m of all strategies.

3. Average development of the game

Since

$$S_n = r_{\xi_n} \cdots r_{\xi_1} = r_1^{\eta_1} \cdots r_m^{\eta_m} = s_n(\eta)\ ,$$

whereas η_i = number of x_i occuring in $(\xi_1,\ldots,\xi_n)$,

S_n depends on $(\xi_1,\ldots,\xi_n)$ just through $\eta=(\eta_1,\ldots,\eta_m)$, which is clearly multinomially distributed, i.e.

$$\Pr\,(\eta=(k_1,\ldots,k_m)) = \frac{n!}{k_1!\ldots k_m!}\,p_1^{\,k_1}\ldots p_m^{\,k_m}\;.$$

Writing $S_n = s_n(B,\eta)$ we can ask for that s_n value as a function of B, which is obtained when we consider the "average development" of the game independently of the strategy. This means merely inserting $E\eta$ instead of η:

$$s_n(B,E\eta) = r_1^{\,np_1}\ldots r_m^{\,np_m} = \exp(n\sum_{i=1}^{m} p_i \log r_i).$$

This criterion leads again to the maximization of f(B).

3. A CONVEX OPTIMIZATION PROBLEM

We now have to solve the following problem of nonlinear programming: maximize

$$f(b_0,b_1,\ldots,b_m)= \Sigma p_i \log(\frac{b_i}{q_i} + b_0) \text{ subject to the constraint}$$

$$b_i = 1,\; b_k\geq 0 \quad k=0,\ldots,m.$$

It is easy to show that the above problem is equivalent to maximizing $g(r_1,\ldots,r_m) = \sum_{i=1}^{n} p_i \log r_i$ subject to the constraint $1\leq\Sigma r_i q_i$, $r_k(\Sigma q_i-1) - \Sigma r_i q_i + 1\geq 0$ $k=1,\ldots,m$. Let Σ'_m denote the domain of R^m defined by these inequalities.

The Hessian matrix with elements

$$\frac{\partial^2 g}{\partial r_i \partial r_j} = 0 \qquad \text{for } i\neq j$$

$$\frac{\partial^2 g}{\partial r_i \partial r_i} = -\frac{p_i}{r_i^{\,2}}$$

is a negativ definite matrix if $p_i>0$ for $i=1,\dots,m$, which we assume from now on. Hence g is strictly concave and g has an unique maximum on Σ'_m.

Since positive linear combinations of concave functions are again concave, f is concave too. But f needs not be strictly concave as the case q=1 shows, where the maximum is obtained along a straight line segment through Σ_m.

Consider first the case $\Sigma q_i = \frac{1}{q} > 1$. Here the transformation $r_i = \frac{b_i}{q_i} + b_0$ from Σ_m to Σ'_m is invertible; the inverse is given by

$$b_0 = \frac{\Sigma r_i q_i - 1}{\Sigma q_i - 1}$$

$$b_k = r_k q_k - \frac{\Sigma r_i q_i - 1}{\Sigma q_i - 1} q_k \qquad k=1,\dots,m.$$

Hence f has a unique maximum on Σ_m.

The case of a zero-sum game, i.e. q=1, is different. Considering Σ_m as a subset of the hyperplane of R^{m+1} given by the normal vector $(1,\dots,1)$ and point under consideration $(1,0,\dots,0)$, it is easy to see that all points on a straight line through Σ_m having direction vector $(-1, q_1,\dots,q_m)$ are mapped into the same point of Σ'_m and thus represent equivalent strategies. Projecting a point $B\in\Sigma_m$ along the straight line of equivalent strategies into the hyperplane $b_0=0$ gives an equivalent strategy E^* where the player stakes his whole capital:

$$b_i^* = b_i + q_i b_0 \quad i=1,\dots,m, \quad b_0^* = 0.$$

B^* is a strategy since q=1. On the other hand for each strategy B with $b_0=0$ there is an equivalent stragegy B^* with maximum money retained:

simply set $b_o^* = \min\{\frac{b_i}{q_i}: i=1,\ldots,m\}$ and $b_i^* = b_i - q_i\, b_o^*$, $i=1,\ldots,m$ and $B^* \in \Sigma_m$ follows again from q=1. B^* is the other boundary point of the straight line segment of equivalent strategies.

If q=1 we can without loss of generality restrict ourselves to the case $b_o=0$. We thus can consider B as a probability distribution on X and

$$f(b_1,\ldots,b_m) = \Sigma p_i \log \frac{p_i}{q_i} - \Sigma p_i \log \frac{p_i}{b_i} = D(P||Q) - D(P||B)$$

whereas D denotes the Kullback-Leibler information number. Thus f is maximized iff we choose B:=P, independent whatever is Q! Clearly $f_{max}=0$ iff P=Q and $f_{max}>0$ else. Thus the game pays whenever $P \neq Q$, that is the other players don't bahave rational.

Turning now to the case q<1, the player will not bet on all outcomes if he follows an optimal strategy. Denote for any strategy B the set of outcomes on which the player bets by

$Z=\{i: b_i>0,\ i=0,\ldots,m\}$ and define $P(Z) = \sum_{\substack{i\in Z\\ i\neq 0}} p_i$;

$Q(Z)$, $Q^*(Z)$ are defined similarly.

Theorem 1.

1. There exists a set Z such that the optimal strategy B for the case q<1 is given by

$$b_k = \begin{cases} 0 & \text{if } k \notin Z \\ p_k - q_k\, b_o > 0 & \text{if } k \in Z \text{ and } k \neq 0 \\ \frac{1-P(Z)}{1-Q(Z)} > 0 & \text{if } k = 0. \end{cases}$$

2. There is a constant λ, $0 \le \lambda \le 1$, such that

$$Z = \{i : \frac{p_i}{q_i} > \lambda\} \cup \{0\}; \text{ in fact } \lambda = b_0 \text{ holds.}$$

3. $Q(Z) \le P(Z)$ holds for Z defined above.

4. $$f_{max} = \sum_{\substack{i \in Z \\ i \neq 0}} p_i \log \frac{p_i}{q_i} + P(Z^c) \log b_0 \ge 0.$$

Proof. The Kuhn Tucker Theorem (1951) implies that the B maximizing f(B) subject to the constraint $B \in \Sigma_m$, satisfies

$$t_k \le \sum_{i=0}^{m} b_i t_i =: t, \; k=0,\dots,m,$$

whereas

$$t_0 := \frac{\partial f}{\partial b_0} = \sum_{i=1}^{m} \frac{p_i q_i}{b_i + q_i b_0} ,$$

$$t_k := \frac{\partial f}{\partial b_k} = \frac{p_k}{b_k + q_k b_0} \quad k \neq 0 .$$

From $\Sigma b_i = 1$, $b_i \ge 0$ follows

$$t_k = t \quad \text{if} \quad b_k > 0, \text{ that is } k \in Z.$$

But we see $t=1$, since

$$t = \sum_{i=1}^{m} b_i t_i + b_0 t_0 = \sum_{i=1}^{m} \frac{b_i p_i}{b_i + q_i b_0} + \sum_{i=1}^{m} \frac{b_0 p_i q_i}{b_i + q_i b_0} = \sum_{i=1}^{m} p_i = 1.$$

Thus $b_k = p_k - q_k b_0$ for $k \in Z$ and $b_k > 0$ by definition.

$$1 = b_0 + \sum_{\substack{k \in Z \\ k \neq 0}} b_k = b_0 + \sum_{\substack{k \in Z \\ k \neq 0}} p_k - b_0 \sum_{\substack{k \in Z \\ k \neq 0}} q_k \quad \text{implies}$$

$$b_0 = \frac{1 - P(Z)}{1 - Q(Z)} .$$

Suppose $0 \notin Z$ that is $b_o = 0$ then the conditions $t_k \le t$ for $k=1,\ldots,m$ imply $\frac{p_k}{b_k} \le 1$ and $b_k = p_k$ follows.

But now $t_o \le t$ gives $\sum_{i=1}^{m} q_i \le 1$, which contradicts the assumption $q<1$.

To establish 2. note that for $k \notin Z$, $k \neq 0$ $\quad t_k = \frac{p_k}{q_k b_o} \le 1$ holds and that for $k \in Z$, $k \neq 0$ $\quad 1 = t_k = \frac{p_k}{b_k + q_k b_o} < \frac{p_k}{q_k b_o}$.

3. follows from $\frac{1-P(Z)}{1-Q(Z)} = b_o \le 1$, 4. follows by computation and $f_{max} \ge f\,(1,0,\ldots,0) = 0$.

Corollary 1.

1. $Z=\{0\}$, that is the optimal strategy is not playing, iff $\frac{p_k}{q_k} \le 1 \quad$ for $k=1,\ldots,m$.
2. Since the optimal strategy is unique it follows from 1. that $f_{max} > 0$ if $Z \neq \{0\}$.

Proof. $Z=\{0\}$ iff $b_o = 1$ and from 2. of the above theorem follows $\frac{p_k}{q_k} \le 1$. If $p_k \le q_k$ then there is no set Z fulfilling $Q(Z) < P(Z)$; therefore we must have $Q(Z) = P(Z)$ in 3. of the above theorem and $b_o = 1$ follows.

Remark 1. It is interesting that the optimal strategy also includes outcomes to bet on, for which $\frac{p_k}{q_k} < 1$ may hold as the following example shows:

X	x_1	x_2	x_3	
P	1/2	1/10	2/5	
Q	2/5	1/9	2/3	$\frac{45}{53} = q$
P/Q	5/4	9/10	3/5	
B	19/110	1/110	0	$\frac{9}{11} = b_0$

Giving an algorithm for the enumeration of Z is crucial for the computation of the optimal strategy. If we renumber the outcomes and put different outcomes with the same $\frac{p_i}{q_i}$ -value together we can arrive at the following inequalities:

$$\frac{p_m}{q_m} < \frac{p_{m-1}}{q_{m-1}} < \ldots < \frac{p_1}{q_1} .$$

Now $Z = \{0,1,\ldots,i\}$ iff $\frac{p_i}{q_i} > \frac{1-P(Z)}{1-Q(Z)} \geq \frac{p_{i+1}}{q_{i+1}}$ which is the same as

$$\frac{q_i}{p_i}\left(1 - \sum_{k=1}^{i-1} p_k\right) + \sum_{k=1}^{i-1} q_k < 1 \leq \frac{q_{i+1}}{p_{i+1}}\left(1 - \sum_{k=1}^{i} p_k\right) + \sum_{k=1}^{i} q_k .$$

We define $\alpha_0=0$, $\alpha_i = \frac{q_i^*}{p_i}\left(1 - \sum_{k=1}^{i} p_k\right) + \sum_{k=1}^{i} q_k^*$ $i=1,\ldots,m$

and it is easily seen that the sequence α_i is strictly increasing from 0 to $1=\alpha_m$. Multiplying the above inequality with q gives the following

Corollary 2.

If $\frac{p_m}{q_m^*} < \frac{p_{m-1}}{q_{m-1}^*} < \ldots < \frac{p_1}{q_1^*}$ holds, then $Z=Z(q) = \{0,\ldots,i\}$

follows, wehre i is that index for which $\alpha_i<q\leq\alpha_{i+1}$ is true, $i=0,1,\ldots,m-1$, i.e.

$$i = \max\{k:\alpha_k<q\} = \min\{k:\alpha_k\geq q\} - 1.$$

If we now consider the cases $q<1$ and $q=1$ simultaneously we obtain the following

Corollary 3.

If either $0\notin Z$ or $Z=\{0,1,\ldots,m\}$ then $q=1$ holds.

Proof. The inequality $Q(Z)\leq P(Z)$ of Theorem 1 rules out the possibility $Z=\{0,1,\ldots,m\}$, since $\sum_{i=1}^{m} q_i = \frac{1}{q} > 1$.

If we let P and Q* fixed and consider the optimal strategy as a function of q it is easily seen that this function is continuous for $0\leq q\leq 1$:

$$b_o(q) = \frac{1-P(Z(q))}{1-\frac{Q^*(Z(q))}{q}}$$

whereas $Z(q)$ has been defined in Corollary 2.

Thus b_o is a decreasing and the b_i's $i=1,\ldots,m$ are increasing functions of q. $f_{max} = f_{max}(q)$ is a continuously differentiable function increasing from 0 to $D(P||Q^*)$ with derivative

$$f'_{max}(q) = \frac{1}{q}\,(1 - b_o(q)).$$

REFERENCES

[1] Bell, R.M. and Cover, T.M. (1980). Competitive optimality of Logarithmic Investment. Math. of Operations Research 5 161-166.

[2] Breiman, L. (1961). Optimal Gambling Systems for Favorable Games. Proc. Fourth Berkeley Symposium on Math. Stat. and Prob. 1, 65-78. University of California Press, Berkeley California.

[3] Cover, T.M. and King, R.C. (1978). A Convergent Gambling Estimate of the Entropy of English. IEEE Trans. Inf. Th. IT-24, 413-421.

[4] Cover, T.M. (1984). An Algorithm for Maximizing Expected Log Investment Return. IEEE Trans. Inf. Th. IT-3 , March 1984.

[5] Kuhn, H. and Tucker, A. (1951). Nonlinear Programming. Proc. Second Berkeley Symposium on Math. Stat. and Prob. 481-492, University of California Press, Berkeley, Calif.

[6] Nemetz, T. and Simon, J. (1978). On Estimating the Entropy of Written Hungarian by Gambling Technique. Trans. of Eighth Prague Conf. on Inf. Theory, Vol. B. 69-76.

Johannes Adler
Department of Statistics and Probability Theory
Technical University Vienna
Argentinerstraße 8/7.
A-1040. Wien

Proc. of the 4th Pannonian Symp. on Math. Stat., Bad Tatzmannsdorf, Austria 1983
W. Grossmann, G. Pflug, I. Vincze, W. Wertz, eds

REALISTIC AND ABSTRACT ROULETTE, A COMPARISON

J.H. AHRENS AND U. DIETER

Graz

SUMMARY

Existing roulette theory teaches that a gambler who is allowed to stake arbitrary amounts of money in order to attain a predetermined goal should employ bold strategies and prefer betting on single holes. We supply a simple algorithm for calculating bold winning probabilities and compare these with their counterparts in realistic roulette where all stakes must be integer multiples of a minimum bet. There the winning probabilities come close to the theoretical bounds if the maximum permissible stakes of the casino are large enough, but optimal betting policies are rather different from bold one-hole playing.

Key word and phrases: Gambling, optimal strategies, bold strategies, ruin problems.

1. INTRODUCTION

At a roulette table with m numbers on the layout a player who covers n of these with his bet wins with probability n P where P=1/m: the m numbers are matched by the m holes on the roulette wheel in which the ball will come to rest with equal probabilities 1/m, always assuming that the machine is in perfect working condition. Most casioes in Europe offer m=37 numbers from 0 to 36, whereas establishments in America tend to include a double zero 00 resulting in m=38. If roulette were a fair game, a winning player who has staked X units of money on a set of n holes should be payed X/nP units including the returned stake. However, casinoes are 'subfair', and the amount returned to the winner is only X/nR where R P. Hence, for every dollar staked the expected gain of the casino is 1-nP/nR=1-P/R dollars.

The large casinoes all have R=1/36 giving an advantage or '*vigorish*' of 1-36/38=1/19=5.26316% if a 00 is present, but only 1-36/37=1/37=2.70270% for singlezero houses. In casinoes with 'prison rule' on even money chances the advantage of the bank is nearly halved if the gambler choses n=18, like red - see Sagan (1981). There are some other roulette type games, for instance slot machines with numbers 0 to 12. For these P=1/13, R=1/12, and the vigorish becomes 1/13=7.69231% (see Dieter (1981b)).

In the next section we describe briefly the existing theory of an idealized roulette game in which stakes of *arbitrary* size are permitted. We also supply a feasible *algorithm* for calculating the maximum probabilities of eventually reaching a given target. These are upper bounds for the probabilities of ultimately attaining the goal G from an initial fortune F in completely realistic roulette (F, G integers).

This real-life case is studied in Section 3. It will be seen that *uniform* stakes (no two different n-hole combinations used simultaneously) give results which fall short of the theoretical bounds by at most 1/450 at G=100 and 1/1100 at G=1000 in Las Vegas roulette. This compares favorably with optimum red-black gambling whose shortcomings can be as high as about 1/50 independently of G. It will also be demonstrated that more complicated betting schemes beyond playing offer only marginal improvements on the winning probabilities. The article is concluded with some remarks on European roulette and doubling strategies.

2. ARBITRARY STAKES

In roulette theory the real game is often idealized by assuming that the gamblers are able to stake arbitrary amounts of money, not just integer multiples of the minimum bet allowed at the table. How then should somebody with an infinitely divisible fortune F play so as to maximize the probability of ultimately walking out with a prespecified larger balance G? The first answer is given by L.E. Dubins and L.J. Savage (1965) for *primitive* roulette in which n is fixed. This includes betting on, say, an even money chance only (n=18), or always putting the money on a single hole (n=1). In all of these separate cases one should play *boldly:* the gambler behaves optimally if he always stakes as much as necessary to reach the target G after *one* lucky spin of the wheel, but if his capital is too small for this purpose, he simply risks everything he has left. Infinite divisibility allows a standardization of the problem: $f=F/G$ $(0<f<1)$ is to be increased to 1 with maximum possible probability.

So let p (=nP) and r (=nR) be the parameters of a primitive subfair $(0<p<r<1)$ casino. A stake x is increased to x/r on winning (with probability p) and lost with probability $\bar{p}=1-p$. Then the bold strategy of someone trying to reach 1 starting with a forutne f is described as follows ($\bar{r}=1-r$):

<u>Bold Strategy.</u> If $f \leq r$ stake $x=f$. Then f grows to f/r with probability p or goes down to zero with probability $\bar{p}$. If $f \geq r$ stake $x=(1-f)r/\bar{r}$. Then f reaches the target $f+x/r-x=1$ with probability p or it decreases to $f-x=(f-r)/\bar{r}$ with probability $\bar{p}$.

Hence the probabilities $b_{p,r}(f)$ of eventually reaching under the bold strategy are described by

$$b_{p,r}(f) = p\, b_{p,r}(f/r) \qquad \text{if } 0 \le f \le r \qquad (1)$$

$$b_{p,r}(f) = p+\bar{p}b_{p,r}((f-r)/\bar{r}) \qquad \text{if } r \le f \le 1 \qquad (2)$$

With the obvious conditions $b_{p,r}(0)=0$ and $b_{p,r}(1)=1$ these identities completely determine $b_{p,r}(f)$; even more, they translate into the following numerically well-behaved procedure for calculating any $b_{p,r}(f)$.

<u>Algorithm B.</u> Input p, $\bar{p}=1-p$, r, $\bar{r}=1-r$, f. Output $b=b_{p,r}(f)$

1. Initialize $b \leftarrow 0$, $h \leftarrow p/\bar{p}$, and set a tolerance $\delta \leftarrow \varepsilon h$, ($\varepsilon=10^{-d}$ for d-digit accuracy).
2. If $f<r$ set $h \leftarrow hp$ and $f \leftarrow f/r$; otherwise set $h \leftarrow h\bar{p}$, $b \leftarrow b+h$ and $f \leftarrow (f-r)/\bar{r}$.
3. If $h>\delta$ go back to step 2, otherwise return b.

We prove that $0 \le b_{p,r}(f)-b \le \varepsilon$ from the following lemma.

<u>Lemma.</u> After $k+\bar{k}$ iterations of Step 2 in Algorithm B - k cases $f<r$ and $\bar{k}$ cases $f \ge r$ - the $(k+\bar{k})$-th approximation $b_{k+\bar{k}}$ in b obeys $0 \le b_{p,r}(f)-b \le p^k\bar{p}^{\bar{k}}$.

<u>Proof.</u> If $k+\bar{k}=0$ then $b=0$ (Step 1), but $0\ b_{p,r}(f) \le 1$. Hence $0 \le b_{p,r}(f)-b_0 \le p^0\bar{p}^0=1$. For the induction step from $k+\bar{k}-1$ to $k+\bar{k}$ consider the two cases of Step 2.

(1) $f<r$ (for the initial f). Then the first execution of Step 2 leaves b at 0. From $f \leftarrow f/r$ the remaining $k+\bar{k}-1=(k-1)+\bar{k}$ steps may be considered as an $(k+\bar{k}-1)$-step algorithm for $b_{p,r}(f/r)$. However, since the initialization $h \leftarrow p/\bar{p}$ (Step 1) was changed by $h \leftarrow hp$, the resulting approximation to $b_{p,r}(f/r)$ is $b_{k,\bar{k}}/p$. By induction we have $0 \le b_{p,r}(f/r)-b_{k+\bar{k}}/p \le p^{k-1}\bar{p}^{\bar{k}}$, and multiplication by p and formula (1) yield $0 \le b_{p,r}(f)-b_{k+\bar{k}} \le p^k\bar{p}^{\bar{k}}$.

(2) $f \ge r$ (for the initial f). Then Steps 1 and 2 yield $h \leftarrow h\bar{p}=(p/\bar{p})\bar{p}=p$, $b \leftarrow 0+h=p$ and $f \leftarrow (f-r)/\bar{r}$. The remaining

$k+\bar{k}-1=k+(\bar{k}-1)$ steps contribute $b_{k+\bar{k}}$ - p and, from $h \leftarrow h\bar{p}$, $b_{p,r}((f-r)/\bar{r})$ is approximated by $(b_{k+\bar{k}} - b)/\bar{p}$. Now, by induction $0 \leq b_{p,r}((f-r)/\bar{r})-(b_{k+\bar{k}} - b)/\bar{p}\ p^{k}\bar{p}^{\bar{k}-1}$, and after multiplication by $\bar{p}$ formula (2) yields $0 \leq b_{p,r}(f)-b_{k+\bar{k}} \leq p^{k}\bar{p}^{\bar{k}}$.

The error bound ε in Algorithm B follows easily from the lemma. When b is returned in Step 3 after $k+\bar{k}$ steps, h has fallen below δ. But h was initialized to $p/\bar{p}$ and multiplied k times by p and $\bar{k}$ times by $\bar{p}$ in Step 2. Hence $h=(p/\bar{p})p^{k}\bar{p}^{\bar{k}}\ \delta=\varepsilon p/\bar{p}$ (Step 1), and the error bound $p^{k}\bar{p}^{\bar{k}}$ of the lemma is then below ε.

If $b_{p,r}(f)=b$ then $f=b_{r,p}(b)$ (see Dubins and Savage (1965)). Therefore, with the interchanges $p \leftrightarrow r$, $\bar{p} \leftrightarrow \bar{r}$, and $f \leftrightarrow b$ Algorithm B will also calculate the inverse function $b_{p,r}^{-1}$ answering the question: how large does the starting capital f have to be in order to ultimately reach the target 1 with a given probability.

The proof of optimality for the bold strategy is based on

$$p\ b_{p,r}(a)+\bar{p}\ b_{p,r}(c) \leq b_{p,r}(ra+\bar{r}c) \quad \text{if } 0 \leq c \leq a \leq 1 \qquad (3)$$

and the related superadditivity

$$b_{p,r}(a)+b_{p,r}(c) \leq b_{p,r}(a+c) \qquad \text{if } a+c \leq 1 \qquad (4)$$

$$b_{p,r}(a)+b_{p,r}(c) \leq b_{p,r}(a+c-1)+1 \qquad \text{if } a+c \geq 1. \qquad (5)$$

In the special case of red-black casinoes ($r=\bar{r}=0.5$) the inequality (3) is not hard to verify, but the case of general r is very difficult. For details compare the book of Dubins and Savage (1965), Dhapter 6.3.

Formula (4) applied n-1 times to $b_{p,r}(n\ f)=b_{p,r}(f+f+...+f)$ yields

$$n\ b_{p,r}(f) \leq b_{p,r}(nf) \qquad \text{if } n>0 \text{ integer and } n\ f \leq 1. \tag{6}$$

From (6) it follows rather easily that a player who has the choice of different primitive roulettes should always prefer the one-hole (n=1) game to betting on multiple (n>1) combinations:

$$b_{nP,nR}(f) \leq b_{P,R}(f) \qquad \text{if } n>0 \text{ integer and } 0 \leq f \leq 1. \tag{7}$$

Proof. The special case f=nP is derived from (6) using $b_{p,r}(r)=p$ (see (2)).

$$b_{nP,nR}(nR)=nP=n\ b_{P,R}(R) \leq b_{P,R}(nR) \tag{8}$$

Generally, an n-hole gambler A_n who stakes x units of a capital f reaches $f^+=f-x+x/nR$ with probability nP or falls back to $f^-=f-x$ with probability 1-nP. This is equivalent to a game in which A_n starts at f =nR and reaches 1 with probability nP and 0 with probability 1-nP. So A_1 may play this transformed game with his 1-hole bold strategy offering him a change $b_{P,R}(nR) \geq nP$ of reaching 1. Therefore A_1 can outperform A_n in respect to every move of A_n; this proves (7).

Algorithm B is easy to program in computer languages, even in pocket calculator codes. Our many numerical results indicate that (6), (7) and (8) may not generally hold for any non-integer n; but if n is large, counterexamples require rather small p and r. In Table 1 we compare n-hole players who are out to double their fortunes (f=0.5) in American and European roulette. Note that the winning probabilities for n=12 are smaller than the ones for n=18. This does not contradict (8) since $\frac{18}{12}$ = 1.5 is not an integer. (See Dubins and Savage (1965),

Chapter 9, Ex. 1 for similar examples.)

In casinoes with 3-prison rule n=18 can be better than even n=1. This is demonstrated in the last column of Table 1 for f= .5. However, the advantage of red-black chances disappears when f is small. Using Algorithm B it was calculated that for f<.435282 (American) or f<.432976 (European roulette) the one-hole probabilities are superior again.

The first column (n→0) of Table contains

$$b_{P/R}(f)=\lim_{n\to 0} b_{nP,nR}(f)=1-(1-F)^{P/R}, \qquad (9)$$

a limit which may be reached to any degree of accuracy by a gambler who is allowed to stake any amount of money on arbitrarily small fractions of one hole. This bound (9) is stated in [3], Chapter 9.3; it follows quickly from (2) which may be written as

$$b_{NP,nR}(f)-b_{nP,nR}(f-h)=nP(1-b_{nP,nR}(f-h)),\ h=(1-f)\ \frac{nR}{1-nR}.$$

After a division by h we obtain for n→0 (and hence h→0)

$$\frac{d}{df}\ b_{P/R}(f)=\frac{P}{R}\ \frac{1-b_{P/R}(f)}{1-f} \text{ or } \frac{d}{df}\ \ln(1-b_{P/R}(f))= -\frac{P}{R}\ \frac{1}{1-f}.$$

With the initial condition $b_{P/R}(0)=0$ this differential equation has the solution (9).

In reality players have to use at least one hole, and G.J. Smith (1967) and L.E. Dubins (1968) demonstrated that the 1-hole probabilities $b_{P,R}(f)$ are in fact universal bounds in the sense that nobody can do better even if all comtinations of stakes on different numbers are considered.

Still, the assumption of an infinitely divisible fortune is not valid either in existing casinoes. In the next section we shall try to find out how closely a gambler may approach the bound $b_{P,R}(f)$ in realistic roulette.

Table 1. n-hole winning probabilities $b_{nP,n/36}(0.5)$ for doubling divisible

number n of holes	n→0	1	2	3	4	6	12	18	18 3 prisons
American Roulette P = 1/38	.48142	.48089	.48070	.48032	.47947	.47795	.47230	.47368	.48612
European Roulette P = 1/37	.49054	.49027	.49017	.48998	.48954	.48876	.48584	.48649	.49306

3. REALISTIC STAKES

In existing roulette stakes are always integer multiples of a minimum bet which we ase as our monetary unit. Hence only integer amounts may be placed on the following n-hole combinations: n = 1,2,3,4,6,12, and 18. We can safely ignore then n=5 bet in some American casinoes because of its excessive vigorish of 3/38. However, further n'-hole bets are possible by placing identical amounts on several n-hole combinations. For instance, 8 holes are covered by two non-overlapping 4-hole combinations. Generally, a stake X on n holes is realistic if X is an integer multiple of M_n according to the following list:

n	1	2	3	4	5	6	7	8	9	10	11	12	13	14	15	16	17	18
M_n	1	1	1	1	5	1	7	2	3	5	11	1	13	7	5	4	17	1

n	19	20	21	22	23	24	25	26	27	28	29	30	31	32	33	34	35
M_n	-19	5	7	11	23	2	25	13	9	7	29	5	31	8	11	17	35

Here we have also ignored that some houses permit placing one unit on 24 holes. But odd amounts on 24 holes are awkward since the payoff is in half units, and our final results indicate that this possibility will not be part of any optimal strategies.

We assume again that the gambler is trying to increase his fortune F to G>F with maximum probability $w(F\to G)$. If he playes in the style of uniform roulette, he selects only one n for each round and a stake $X=Y\ M_n$ where Y is a positive integer. Then, with probability nP his capital F moves up to $F^+=F-X+X/nR$, otherwise it goes down to $F^-=F-X$. Consequently,

$$w(F\to G)=nP\ w(F^+\to G)+(1-nP)w(F^-\to G), \tag{10}$$

where $F^+=F-X+x/nR$, $F^-=F-X$.

As an example of non-uniform staking consider a capital F=6 and a goal G=100. Now place five dollars onto the n=2 numbers 1,2 (a cheval) and one additional chip onto the n=4 numbers 1,2,4,5 (en caree). If netiher number wins, the capital 6 is lost. If 4 or 5 are called, the fortune moves up to $6-6+1\times 36/4=9$, but if 1 or 2 wins it goes to $6-6+1\times 36/4+5\times 63/2=$ $=99$. Hence

$$w(6\to 100)\geq \frac{2}{38}\ w(99\to 100)\ +\ \frac{2}{38}\ w(9\to 100)\ +\ \frac{34}{38}\ w(0\to 100) \tag{11}$$

in American roulette where P=1/38. This example will be used later on. Generally, all we need to know is that any combination of stakes for one spin of the whell will produce an inequality of the form

$$w(K \to G) \geq \Sigma\lambda_J w(J \to G) \quad \text{where} \quad \Sigma\lambda_J = 1 \quad \text{and} \quad \lambda_J \geq 0. \tag{12}$$

In (12) $\lambda_J \neq 0$ holds only if J is possible new balance after the round. A stake belongs to uniform roulette if exactly two of the λ_J are positive. Therefore we define S_2 as the set of all uniform stakes, and, generally, S_ν as the totality of all staking possibilities which lead to at most ν coefficients $\lambda_J > 0$ in (12). Then the example (11) belongs to $S_3 \backslash S_2$. Every conceivable bet is contained in S_m(S_{38} or S_{37}), but this set is of astronomical size; numerically we were able to cope only with S_2 and S_3 on the basis of the following theorem.

Theorem. Let S be a given set of staking possibilities s, and define the subsets S(F) (0<F<G) of all S-strategies which are playable at F, since their total stakes required do not exceed F. Assume that none of the S(F) is empty. Furthermore, let λ_J^s be the probabilities that the fortune F moves to J after a decision on the s-bet by a spin of the wheel. Then the optimal probabilities $w(F \to G)$ of ultimately attaining the goal G are obtained by solving the system

$$w(0 \to G) = 0, \quad w(F \to G) = \max_{s \in S(F)} \Sigma\lambda_J^s w(J \to G) \quad \text{if} \quad 0 < F < G, \quad w(F \to G) = 1 \quad \text{if} \quad F \geq G. \tag{13}$$

Proof. Let $v(F \to G)$ denote the true optimal probabilities of ultimately reaching G from F under S, and assume that the difference $v(F \to G) - w(F \to G)$ attains its maximum Δ at F=K:

$v(K\to G)-w(K\to G)=\Delta\geq 0$. The optimum $v(K\to G)$ is achieved by some bet $\sigma\in S(K)$ such that $v(K\to G)=\Sigma\lambda_J^{\sigma}v(J\to G)$, whereas

$$w(K\to G)=\max_{s\in S(K)}\Sigma\lambda_J^{s}w(J\to G)=\Sigma\lambda_J^{\tau}w(J\to G) \text{ for some } \tau\in S(K).$$

Hence

$$v(K\to G)-w(K\to G)=\Sigma\lambda_J^{\sigma}v(J\to G)-\Sigma\lambda_J^{\tau}w(J\to G)\leq\Sigma\lambda_J^{\sigma}(v(J\to G)-w(J\to G)) .$$

But $v(J\to G)-w(J\to G)\leq\Delta$ by assumption, and the λ_J^{σ} belong to a convex combination. Therefore $v(J\to G)-w(J\to G)=\Delta$ for all J which occur with positive probabilities λ_J^{σ} one round after the bet σ. Of course, one of these J belongs to a loss and is smaller than K.

Repeating this argument we obtain a sequence of decreasing J for which $v(J\to G)-w(J\to G)=\Delta$, and this must end with $v(0\to G)-w(0\to G)=\Delta$. Hence $\Delta=0$, and the theorem is proved.

The system (13) has a dominant diagonal and can therefore be solved by iteration. After some experimentation we found that the following procedure worked very well for various sets S of betting strategies.

1. Initialize $w(F\to G)\leftarrow 1-(1-\frac{F}{G})^{P/R}$ for $F=0,1,\ldots,G$ (cf (9)) and $w(F\to G)$ 1 for $F=G+1,G+2,\ldots,G+z$ (Z is the largest contemplated overshot; Z=35 is safe). Initialize the optimal strategies $\sigma(F)$ to zero, (i.e. their stakes and hole combinations).

2. For $F=1,2,\ldots,G-1$ replace $w(F\to G)\leftarrow \max_{s\in S(F)}\Sigma\lambda_J^{s}w(J\to G)$, and record the strategies $\sigma(F)$ for which the maxima were attained. (Note that the replacement is immediate: if $J<F$ the new $w(J\to G)$ is used, this proved to be faster.) If no changes of $\sigma(F)$ occur and if all absolute differences between new and old $w(F\to G)$ are below a tolerance δ then print results and exit. (We used a cautious $\delta=10^{-12}$ for a printout of 8 decimals).

3. Keep the strategies $\sigma(F)$ fixed: for $F=1,2,\ldots,G-1$ replace $w(F\to G)\leftarrow\Sigma\,{}^{\sigma(F)}_{J}\,w(J\to G)$. Repeat this step until all changes are below δ or after at most 16 repetitions. Then go back to Step 2.

Our (Siemens 7760) Fortran program has a number of special features. The parameters G, 1/P, 1/R and the number of prisons - if any - for even money chances may be chosen. Maximum bets (house limits) may be imposed on each n-hole combination separately. For instance, by setting these limits to zero for all $n\neq 18$ the red-black probabilities $w(F\to G)=b_{18P,18R}(F/G)$ are recovered, they are not affected by the lack of divisibility. On the other hand, it only n=1 is allowed, $w(F\to G)$ is considerably smaller than $b_{P,R}(F/G)$ since overshots are unavoidable.

The optimal strategies $\sigma(F)$ are not uniquely determined, but a built-in trie braking facility allows to prefer large stakes thereby recovering the true red-black strategies.

Steps 2 and 3 are organized for $S=S_2$ and $S=S_3$. The case S_3 uses the results of S_2 as a starting point. The inclusion of the mathematically redundant Step 3 cuts the computation time by factors 5-10. For G=100, 1/R=36 and 1/P=38 (American roulette) we needed 4 iterations of Step 2, 31 of Step 3 and only 1.7 seconds CPU-time to find optimal S_2-strategies, but the S_3-part added 3 iterations of Step 2, 11 of Step 3 and 123,4 seconds. At G=1000 the S_2-strategies required 3 minutes, and we were not prepared to spend an estimated $3\frac{1}{2}$ hours on the S_3-case.

Many tables for various types of roulette and staking restrictions were produced. For a proper comparison with the theory of Section 2 we display some results in respect to Las Vegas roulette (P=1/38, R=1/36) in Table 2 - unfortunately a shorter list would not show all the characteristic features.

The numbers in brackets are distances ($\Delta\times10^5$) to the first bound $b_{P,R}(F/G)$, the one-hole winning probabilities in the case of infinitely divisible fortunes. The second bound $b_{P/R}(F/G)=$ $=1-(1-F/G)^{P/R}$ concerns arbitrarily small holes (see (9)).

Uniform (S_2) discrete roulette with a target G=100 yields fair approximations to $b_{P,R}(F/G)$ ($\Delta\leq.00222355$) and smaller fluctuations compared with red-black casinoes ($\Delta\leq.01951830$). The optimal $S_.$-strategies are listed. For instance, if someone starts with F=50 units, he should bet X=10 on n=6 holes. If he is lucky, he will reach $F^+=100$ immediately, otherwise he will fall back to $F^-=40$. There he bets X=12 units on n=6 holes and reaches $F^+=100$ or $F^-=28$, etc.

Note that at F=16 the player should overshoot the target 100 by 1, an effect that does not occur when the fortunes are infinitely divisible. The second deviation from the theory in Section 2 is apparent in the S_3-column: non uniform stakes do offer small improvements. Consider once more the example (11) and substitute the S_2-probabilities for $w(99\to100)$ and $w(9\to100)$. Then

$$w(6\to100)\geq\frac{2}{38}\times.98651277 + \frac{2}{38}\times.08430775 + \frac{34}{38}\times0 = .05635897$$

which is larger than $w(6;S_2)=.05571252$. If S_3 strategies are allowed for all F, then $w(6\to100)$ benefits from other improvements, and finally $w(6;S_3)=.05637669$. This example F=6 shows the largest edge of S_3 over S_2 at G=100, non-uniform stakes occured in 33 out of the 99 cases 0<F<100. These 'useful' bets were all

Table 2. Comparisons of winning probabilities

$\frac{F}{G}$	RED - BLACK $p=18/38$, $r=1/2$		UNIFORM STAKES (SET S_2) $G = 100$, $F = Gf = 100f$						SET S_3 $G = 100$	SET S_2 $G = 1000$		BOUNDS		
(=f)	$b_{p,r}(f)$		$w(F \to G)$		F^-	X	n	F^+	$w(F \to G)$	$w(F \to G)$		$b_{P,R}(f)$	$1-(1-f)^{P/R}$	
.01	.00683	(223)	.00903	(3)	0	1	1	36	.00903	.00905	(1)	.00906	.00948	(-42)
.02	.01443	(400)	.01837	(6)	0	2	1	72	.01838	.01841	(2)	.01843	.01896	(-53)
.03	.02261	(570)	.02743	(88)	1	2	1	73	.02778	.02812	(19)	.02831	.02844	(-13)
.04	.03046	(699)	.03679	(66)	2	2	1	74	.03700	.03732	(13)	.03745	.03794	(-49)
.05	.03881	(815)	.04659	(37)	0	5	2	90	.04660	.0 690	(6)	.04696	.04743	(-47)
.06	.04774	(891)	.05571	(94)	1	5	2	91	.05638	.05653	(12)	.05665	.05693	(-28)
.07	.05598	(992)	.06514	(76)	2	5	2	92	.06566	.06579	(11)	.06590	.06644	(-54)
.08	.06430	(1141)	.07510	(61)	0	8	3	96	.07511	.07565	(6)	.07571	.07595	(-24)
.09	.07310	(1194)	.08431	(73)	1	8	3	97	.08463	.08498	(6)	.08504	.08547	(-43)
.10	.08193	(1253)	.09385	(61)	2	8	3	98	.09403	.09430	(16)	.09446	.09500	(-54)
.11	.09126	(1302)	.10384	(44)	0	11	4	99	.10385	.03409	(19)	.10428	.10452	(-24)
.12	.10079	(1273)	.11334	(18)	1	11	4	100	.11334	.11337	(15)	.11352	.11406	(-54)
.13	.10988	(1341)	.12246	(83)	12	1	1	48	.12247	.12306	(23)	.12329	.12360	(-31)
.14	.11819	(1451)	.13198	(72)	12	2	1	84	.13209	.13249	(21)	.13270	.13315	(-45)
.15	.12671	(1546)	.14154	(63)	10	5	2	100	.14171	.14197	(20)	.14217	.14270	(-53)
.16	.13575	(1619)	.15101	(93)	11	5	2	101	.15101	.15175	(19)	.15194	.15226	(-32)
.17	.14476	(1650)	.16027	(99)	15	2	1	87	.16045	.16106	(20)	.16126	.16182	(-56)
.18	.15432	(1689)	.16986	(135)	11	7	3	95	.17007	.17090	(31)	.17121	.17139	(-18)
.19	.16412	(1632)	.17950	(94)	12	7	3	96	.17956	.18027	(17)	.18044	.18097	(-53)
.20	.17297	(1720)	.18923	(94)	10	10	4	100	.18939	.18981	(36)	.19017	.19055	(-38)

Table 2. Comparisons of winning probabilities (continued).

$\frac{F}{G}$ (=f)	RED - BLACK p=18/38, r=1/2 $b_{p,r}(f)$	UNIFORM STAKES (SET S_2) G = 100, F = GF = 100 f w(F→G)	F^-	X	n	F^+	SET S_3 G = 100 w(F→G)	SET S_2 G = 1000 w(F→G)	BOUNDS $b_{P,R}(f)$	$1-(1-f)^{P/R}$
.21	.18247 (1720)	.19849 (118)	14	7	3	98	.19872	.19942 (25)	.19967	.20014 (-47)
.22	.19266 (1657)	.10825 (98)	15	7	3	99	.20841	.20888 (35)	.20923	.20973 (-50)
.23	.20232 (1660)	.21804 (88)	16	7	3	100	.21804	.21871 (21)	.21892	.21933 (-41)
.24	.21277 (1562)	.22724 (115)	23	1	1	59	.22737	.22803 (36)	.22839	.22894 (-55)
.25	.22438 (1381)	.13704 (115)	23	2	1	95	.23707	.23788 (31)	.23819	.23856 (-37)
.26	.23197 (1565)	.24629 (133)	24	2	1	96	.24661	.24728 (34)	.24762	.24818 (-56)
.27	.24041 (1706)	.25613 (134)	25	2	1	97	.25617	.25712 (35)	.25747	.25781 (-34)
.28	.24950 (1738)	.26587 (101)	19	9	4	100	.16592	.26656 (32)	.26688	.26744 (-56)
.29	.25822 (1854)	.27535 (141)	27	2	1	99	.27546	.27632 (44)	.27676	.27709 (-33)
.30	.26750 (1867)	.28519 (98)	28	2	1	100	.28524	.28586 (31)	.28617	.28674 (-57)
.31	.27742 (1862)	.29457 (147)	27	4	2	99	.29476	.29553 (51)	.2960	.29639 (-35)
.32	.28658 (892)	.30451 (99)	28	4	2	100	.30456	.30516 (34)	.30550	.30606 (-56)
.33	.29582 (1952)	.31381 (153)	32	1	1	68	.31408	.31493 (41)	.31534	.31573 (-39)
.34	.30560 (1926)	.32382 (104)	28	6	3	100	.32388	.32446 (40)	.32486	.32541 (-55)
.35	.31542 (1923)	.33326 (139)	22	13	6	100	.33339	.33419 (46)	.33465	.33509 (-44)
.36	.32578 (1852)	.34314 (116)	28	8	4	100	.34320	.34376 (54)	.34430	.34479 (-49)
.37	.33636 (1763)	.35249 (150)	36	1	1	72	.35273	.35364 (35)	.35399	.35449 (-50)
.38	.34647 (1741)	.36246 (142)	28	10	5	100	.36252	.36310 (78)	.36388	.36420 (-32)
.39	.35569 (1769)	.37183 (155)	38	1	1	74	.37205	.37295 (43)	.37338	.37392 (-54)
.40	.36547 (2815)	.38178 (154)	28	12	6	100	.38187	.38265 (67)	.38332	.38365 (-33)

Table 2. Comparisons of winning probabilities (continued).

$\frac{F}{G}$ (=f)	RED - BLACK p=18/38, r=1/2 $g_{p,r}(f)$	UNIFORM STAKES (SET S_2) G = 100, F = GF = 100 f w(F→G)	F^-	X	n	F^+	SET S_3 G = 100 w(F→G)	SET S_2 G = 1000 w(F→G)	BOUNDS $b_{P,R}(f)$	$1-(1-f)^{P/R}$
.41	.37521 (1763)	.39118 (166)	40	1	1	76	.39147	.39233 (51)	.39284	.39399 (-55)
.42	.38522 (1744)	.40110 (156)	28	14	7	100	.40117	.40217 (49)	.40266	.40313 (-47)
.43	.39584 (1668)	.41087 (165)	36	7	4	99	.41092	.41179 (73)	.41252	.41288 (-36)
.44	.40673 (1537)	.42065 (145)	37	7	4	100	.42086	.42169 (41)	.42210	.42265 (-55)
.45	.41657 (1545)	.43059 (143)	34	11	6	100	.43067	.43139 (63)	.43202	.43202 (-40)
.46	.42712 (1462)	.44006 (168)	5	1	1	81	.44014	.44105 (69)	.44174	.44220 (-46)
.47	.43844 (1302)	.44979 (467)	4	3	2	98	.45000	.45103 (43)	.45146	.45199 (-53)
.48	.44918 (1221)	.45985 (154)	5	3	2	99	.45992	.46063 (76)	.46139	.46179 (-40)
.49	.46079 (4038)	.46953 (164)	6	3	2	100	.46960	.47047 (70)	.47117	.47160 (-43)
.50	.47368 (4721)	.47940 (149)	0	10	6	100	.47947	.48033 (56)	.48089	.48142 (-53)
.51	.48128 (952)	.48911 (169)	45	6	4	99	.48918	.48918 (65)	.49080	.49125 (-45)
.52	.48972 (1108)	.49900 (180)	46	6	4	100	.49907	.49989 (91)	.50080	.50109 (-29)
.53	.49881 (1164)	.50860 (185)	52	1	1	88	.50867	.50982 (63)	.51045	.51095 (-50)
.54	.50753 (1276)	.51847 (182)	50	4	3	98	.51857	.51976 (53)	.52029	.52081 (-52)
.55	.51681 (1342)	.528 7 (176)	46	9	6	100	.52856	.52948 (75)	.53023	.53068 (-45)
.56	.52673 (1348)	.53855 (166)	52	4	3	100	.53862	.53938 (83)	.54020	.54057 (-36)
.57	.53589 (1422)	.54825 (186)	56	1	1	92	.54831	.54927 (84)	.55011	.55047 (-36)
.58	.54513 (1477)	.55797 (193)	57	1	1	93	.55807	.55930 (60)	.55990	.56038 (-48)
.59	.55490 (1490)	.56774 (206)	54	5	4	99	.56813	.56914 (66)	.56980	.57030 (-50)
.60	.56472 (1503)	.57810 (165)	52	8	6	100	.57818	.57901 (74)	.57975	.58024 (-49)

Table 2. Comparisons of winning probabilities continued.

$\frac{F}{G}$ (=f)	RED - BLACK p=18/38, r=1/2 $g_{p,r}(f)$	UNIFORM STAKES (SET S_2) G = 100, F = GF = 100 f w(F→G)	F^-	X	n	F^+	SET S_3 G = 100 w(F→G)	SET S_2 G = 1000 w(F→G)	B O U N D S $b_{P,R}(f)$	$1-(1-f)^{P/R}$
.61	.57508 (1464)	.58793 (179)	60	1	1	96	.58801	.58891 (81)	.58972	.59019 (-47)
.62	.58567 (1404)	.59778 (193)	61	1	1	97	.59786	.59888 (83)	.59971	.60015 (-44)
.63	.59577 (1394)	.60769 (202)	62	1	1	98	.60777	.60889 (82)	.60971	.61012 (-41)
.64	.60500 (1471)	.64766 (205)	63	1	1	99	.61779	.61890 (81)	.61971	.62011 (-40)
.65	.61447 (1525)	.62777 (195)	58	7	6	100	.62785	.62893 (79)	.62972	.63012 (-40)
.66	.62452 (1521)	.63779 (194)	64	2	2	100	.63790	.63895 (78)	.63973	.64014 (-41)
.67	.63452 (1523)	.64785 (190)	64	3	3	100	.64796	.64898 (77)	.64975	.65017 (-42)
.68	.64515 (1464)	.65791 (188)	64	4	4	100	.65802	.65901 (78)	.65979	.66022 (-43)
.69	.65603 (1381)	.66797 (187)	64	5	5	100	.66808	.66905 (79)	.66984	.67029 (-45)
.70	.66588 (1406)	.67803 (191)	64	6	6	100	.67815	.67914 (80)	.67994	.68037 (-43)
.71	.67643 (1370)	.68809 (204)	64	7	7	100	.68820	.68932 (81)	.69013	.69048 (-35)
.72	.68775 (1250)	.69815 (210)	64	8	8	100	.69832	.69952 (73)	.70025	.70060 (-35)
.73	.69849 (1184)	.70822 (211)	64	9	9	100	.70845	.70958 (75)	.71033	.71074 (-41)
.74	.71009 (1041)	.71828 (222)	64	10	10	100	.71849	.71970 (80)	.72050	.72090 (-40)
.75	.72299 (4775)	.72887 (187)	70	5	6	100	.72897	.72999 (75)	.73074	.73108 (-34)
.76	.73143 (945)	.73893 (195)	73	3	4	100	.73914	.74016 (72)	.74088	.74128 (-40)
.77	.74080 (1039)	.74921 (198)	75	2	3	99	.74931	.75037 (82)	.75119	.75150 (-31)
.78	.75091 (1046)	.75954 (183)	76	2	3	100	.75974	.76067 (70)	.76137	.76175 (-38)
.79	.76060 (1107)	.76964 (203)	73	6	8	100	.76985	.77102 (65)	.77167	.77202 (-35)
.80	.77091 (1112)	.78015 (188)	76	4	6	100	.78033	.78122 (81)	.78203	.78232 (-29)

Table 2. Comparisons of winning probabilities (continued).

$\frac{F}{G}$ (=f)	RED - BLACK p=18/38, r=1/2 $g_{p,r}(f)$	UNIFORM STAKES (SET S_2) G = 100, F = GF = 100 f w(F→G)	F^-	X	n	F^+	SET S_3 G = 100 w(F→G)	SET S_2 G = 1000 w(F→G)	BOUNDS $b_{P,R}(f)$	$1-(1-f)^{P/R}$
.81	.78193 (1042)	.79037 (198)	80	1	2	198	.79054	.79156 (79)	.79235	.79265 (-30)
.82	.79211 (1057)	.80076 (192)	76	6	9	100	.80092	.80196 (71)	.80168	.80300 (-32)
.83	.80238 (1070)	.81125 (183)	82	1	2	100	.81140	.81238 (70)	.81308	.81338 (-30)
.84	.81323 (1033)	.82174 (182)	82	2	4	100	.82188	.82281 (75)	.82356	.82380 (-24)
.85	.82415 (981)	.83222 (174)	82	3	6	100	.83236	.83323 (73)	.83396	.83425 (-29)
.86	.83566 (879)	.84271 (174)	82	4	8	100	.84283	.84379 (66)	.84445	.84474 (-29)
.87	.84742 (757)	.85319 (180)	82	5	10	100	.85336	.85429 (70)	.85499	.85526 (-27)
.88	.85865 (694)	.86372 (187)	87	1	3	99	.86391	.86492 (67)	.86559	.86583 (-24)
.89	.86890 (730)	.87448 (172)	88	1	3	100	.87465	.87558 (62)	.87620	.87645 (-25)
.90	.87942 (745)	.88524 (163)	88	2	6	100	.88540	.88625 (62)	.88687	.88712 (-25)
.91	.89058 (704)	.89600 (162)	88	3	9	100	.89614	.89703 (59)	.89762	.89784 (-22)
.92	.90170 (671)	.90694 (147)	91	1	4	100	.90707	.90786 (55)	.90841	.90863 (-22)
.93	.91350 (580)	.91789 (141)	91	2	8	100	.91801	.91876 (54)	.91930	.91948 (-18)
.94	.92560 (466)	.92884 (142)	91	3	12	100	.92894	.92974 (52)	.93026	.93042 (-16)
.95	.93654 (477)	.94008 (123)	94	1	6	100	.94016	.94083 (48)	.94131	.94146 (-15)
.96	.94826 (423)	.95131 (118)	94	2	12	100	.95138	.95205 (44)	.95249	.95262 (-13)
.97	.96084 (297)	.96255 (126)	94	3	18	100	.96260	.96341 (40)	.96381	.96392 (-11)
.98	.97277 (258)	.97437 (98)	96	2	18	100	.97441	.97501 (34)	.97535	.97543 (-8)
.99	.98567 (154)	.98651 (70)	98	1	18	100	.98653	.98696 (25)	.98721	.98726 (-5)

of the same kind: stake X_1 onto a set M_1 of n_1 holes and X_2 onto a subset M_2 M_1 of n_2 holes. The program had to try out in vain the two other types of non-uniform S_3-stakes (M_1 $M_2=\emptyset$, and the same stake X for M_1 and M_2 with $M_1 \neq M_2$, M_1 $M_2 \neq \emptyset$, but $n_1=n_2$).

Instead of employing less manageable non-uniform bets one could try to split the capital into finer parts: G=1000 is still within the house limits of some better casinoes. The probabilities $w(F \to 1000)$ in Table 2 are quite close to the bound $W_{P,R}(F/1000)$ ($\Delta \leqq .00090403$) although only uniform (S_2) betting is used. The complete results ($0<F<1000$) contain 24 overshots (18 times to 1001, 3 times to 1002, once to 1003, 1004 and 1008). If capitals could be split into even finer coinage we would see the overshots become smaller compared with the target G, one-hole bets would be used more often (except at the end of the table), and the probabilities $w(F \to G)$ would approximate the bounds $b_{P,R}(F/G)$ ever more closely.

Among other types of roulette the European version ($P=1/37$, $R=1/36$) with (three) prisons is the most important one. There the probabilities of gaining even money for stakes on red, black, even, odd, high or low are $887778/1800559>18/37$, since such bets are given another chance whenever a zero falls (Sagan (1981)). Instead of leaving the stake in prison until it finally wins or loses, one may, in case of a zero, also ask for half the money back (partager) which is marginally better.

For both of these options the program found almost the same uniform (S_2) strategies as in American roulette as long as $f=F/G$ was small enough: $f<0.35$ at G=100 and $f<0.36$ at G=1000. For larger f the advantage of the prison rule takes effect, and the gamblers should then essentially revert to the red-black bold strategies. The insignificant accasional exceptions are caused by disturbances feeding back from the improved winning probabilities at small f. For $G\to\infty$ the lower limit for winning red-black bold strategies will gradually move up to f=0.432976 (see Section 2).

Finally we consider the martingale or doubling scheme in roulette: stake 1 unit on red, and double the stake until you win or until you get stuck at the house limit or at the end of your means. In our context this means risking a fortune of $F=2^g-1$ in order to reach $G=2^g$, a gain of one unit. Even in the last possible g-th round the remaining capital is still 2^{g-1}, Hence $f\geq 0.5$ during the entire game, and in European roulette with prisons there is no better alternative to stricking with doubling on red, because this is the bold 18-hole strategy. However, in Las Vegas the optimal uniform (S_2) strategies are different and require computer outputs as in Table 2.

In Table 3 inverses of ruin probabilities are compared. For instance, if one is able to double up to 10 times, the chance of losing 1023 in American roulette is 1 in 613.11= $=(20/38)^{10}$, whereas 1 in 1024 would be fair. Using optimal uniform strategies instead of doubling decreases the ruin probability to 1 in 667.16 which is still rather unattractive.

Table 3. Inverses of ruin probabilities: $1/(1-w(2^g-1 \to 2^g))$

Number of rounds g	1	2	3	4	5	6	7	8	9	10
Target 2^g	2	4	8	16	32	64	128	256	512	1024
European, 3 prisons	1.9726	3.8912	7.6758	15.141	29.868	58.918	116.22	229.26	452.24	892.09
American red-black	1.900	3.6100	6.8590	13.032	24.761	47.046	89.39	169.84	322.69	613.11
American uniform	1.900	3.6100	6.8613	13.180	25.334	48.689	93.61	180.04	346.51	667.16

REFERENCES

DIETER, U. (1981a) Roulette as a ruin game. Procedings of the of the 5th Symposium on Operations Research, Köln 1980, edited by R. Burkard. Methods of Operations Research 41, 75-78.

DIETER, U. (1981b) The classical ruin problem and electronic roulette machines, in Computational Statistics, edited by H. Büning and P. Naeve, p. 55-70, Walter de Gruyter, Berlin-New York.

DIETER, U. and AHRENS, J.H. (1982) The prison rule in roulette, to appear.

DUBINS, L.E. (1968) A simpler proof of Smith's roulette theorem, Ann. Math. Statist. 39, 390-393.

DUBINS, L.E. and SAVAGE, L.J. (1965) How to gamble if you must: Inequalities for stochastic processes. MacGraw-Hill Book Company New York. Second Edition Dover Publications Inc. New York 1976.

JACOBS, K. (1969) Selecta Mathematica I. Springer Verlag Berlin, p. 28-52.

SAGAN, H. (1981) Markov Chains in Monte Calro. Mathematics Magazine 54, 3-10.

SMITH, G. (1967) Optimal strategy at roulette. Z. Wahrscheinlichkeitstheorie Verw. Gebiete 8, 91-100.

Department of Mathematics
University of Kiel,
Olshausenstrasse 40-60
D-2300 Kiel, West-Germany

Institute of Statistics
Technical University
Hamerlinggasse 6
A-8010 Graz, Austria

Proc. of the 4th Pannonian Symp. on Math. Stat., Bad Tatzmannsdorf, Austria 1983
W. Grossmann, G. Pflug, I. Vincze, W. Wertz, eds

A COMPARISON OF METHODS FOR INTERPOLATION IN TIME SERIES

JIŘÍ ANDĚL

Prague

Summary

Consider a time series $\{X_t\}$ with a missing value X_m. Denote by $\hat{X}_m$ and $\tilde{X}_m$ the best linear forward and backward extrapolations of X_m, respectively. Let X_m^* be the best linear combination of $\hat{X}_m$ and $\tilde{X}_m$ for the approximation of X_m and let X_m^o be the best linear interpolation. The residual variances $\lambda_m^2=E(X_m-X_m^*)^2$ and $\tau_m^2=E(X_m-X_m^o)^2$ are compared in the case that $\{X_t\}$ is a stationary or a non-stationary AR(2) process. Explicit formulas for the covariance function of non-stationary AR(2) processes are also derived.

1. Introduction

Let $\{X_t\}$ be a univariate discrete time series with vanishing mean. We shall assume that $EX_t^2<\infty$ for all t. It may happen that only the variables $\{X_t, t\in K\}$ are known, where K is a given subset of integers. The problem is to estimate the variables X_s for $s\notin K$. We shall consider the cases when $X_m,\dots,X_{m+q-1}$ are not known for some $q\geq 1$, whereas the finite or infinite past $\dots,X_{m-2},X_{m-1}$ and future $X_{m+q},X_{m+q+1},\dots$ are given. Then we have the problem of interpolation.

Assume for a while that both the known past and future have infinite many variables and that $\{X_t\}$ is a stationary ARMA process

$$X_t+a_1X_{t-1}+\ldots+a_nX_{t-n} = Y_t+c_1Y_{t-1}+\ldots+c_rY_{t-r},$$

where $\{Y_t\}$ is a white noise, i.e. uncorrelated random variable with vanishing mean and the same variance σ^2. Let the coefficien a_j, c_k be known. Denote by B the back-shift operator and put

$$A(B) = 1+a_1B+\ldots+a_nB^n, \qquad C(B) = 1+c_1B+\ldots+c_rB^r.$$

If the model is invertible (i.e. if $C(z)\neq 0$ for $|z|\leq 1$), then

$$Y_t = [A(B)/C(B)]X_t = \pi(B)X_t = X_t-\pi_1X_{t-1}-\pi_2X_{t-2}-\ldots$$

Define

$$P(B) = \pi(B^{-1})\pi(B) = \sum_{k=-\infty}^{\infty} p_kB^k.$$

Obviously,

$$p_k = p_{-k} = \sum_{j=0}^{\infty} \pi_j\pi_{j+k} \quad \text{for} \quad k\geq 0.$$

Denote by X_s^o the best linear interpolation of X_s for $s\notin K$ and put

$$\xi_t = \begin{cases} X_t & \text{for } t\in K, \\ X_t^o & \text{for } t\notin K. \end{cases}$$

Then X_s^o can be calculated from the equations

$$\sum_{k=-\infty}^{\infty} p_k\xi_{s+k} = 0, \qquad s\notin K$$

[see Brubacher and Wilson (1976)]. If X_m is the only unknown variable, then

$$X_m^o = -p_o^{-1} \sum_{k\neq 0} p_k X_{m+k} .$$

If $X_m, \ldots, X_{m+q-1}$ are not known, it is necessary to solve the system of linear equations

$$\sum_{j=0}^{q-1} p_{s-j} X_{m+j}^o = - \sum_{\substack{j<0 \\ j\geq q}} p_{s-j} X_{m+j}, \quad s=0,1,\ldots,q-1.$$

In practice, however, the computer programs for extrapolation are frequently used instead of calculating the best linear interpolation. It is possible to use the extrapolation from the previous values (forward extrapolation) as well as from the future values (backward extrapolation). The backward extrapolation is constructed by reversing the time series. Damsleth (1980) proposed to interpolate missing values by the best linear combination of the forward and backward extrapolations. The numerical comparisons of the residual variances show that the reduction in Damsleth's method is not very great for stationary series, but it may be substantial when the series is non-stationary. The aim of our paper is to compare the residual variances for forward extrapolation, backward extrapolation, Damsleth's method and the best linear interpolation, when the investigated process is autoregressive. We include also the non-stationary AR models. Because the formulas become rather complicated, we consider only the case when one variable is missing.

2. The autoregressive model and its covariance function

2.1. The model

Let $\{Y_t\}$ be a white noise with $EY_t=0$, $\operatorname{Var} Y_t=\sigma^2$ and let $X_1,\dots,X_n$ be random variables with finite second moments and vanishing means, which are independent of $\{Y_t\}$. If $a_1,\dots,a_n$ are given numbers, we define $X_{n+1},X_{n+2},\dots$ by the relation

$$X_t+a_1X_{t-1}+\dots+a_nX_{t-n}=Y_t, \qquad t>n. \tag{2.1}$$

Put

$$M=\left\|\begin{matrix} 0 & 1 & 0 & \dots & 0\\ 0 & 0 & 1 & \dots & 0\\ \cdot & \cdot & \cdot & \cdot & \cdot\\ 0 & 0 & 0 & \dots & 1\\ -a_n & -a_{n-1} & -a_{n-2} & \dots & -a_1\end{matrix}\right\|, \quad J=\left\|\begin{matrix} 0 & 0 & \dots & 0 & 0\\ 0 & 0 & \dots & 0 & 0\\ \cdot & \cdot & \cdot & \cdot & \cdot\\ 0 & 0 & \dots & 0 & 0\\ 0 & 0 & \dots & 0 & 1\end{matrix}\right\|,$$

$a_0=1$, $\Lambda=\sigma^2J$, $B_1=\operatorname{Var}(X_1,\dots,X_n)$.

It is well known [Anděl (1976), pp. 164-169] that if

$$\sum_{k=0}^{n} a_kz^k \neq 0 \quad \text{for } |z|\le 1 \tag{2.2}$$

then the matrix equation $B^*=MB^*M'+\Lambda$ has a unique solution B^*. This solution is symmetric and positive definite and it is given by

$$B^* = \sum_{k=0}^{\infty} M^k\Lambda M'^k .$$

The process $\{X_t\}$ is stationary if and only if (2.2) holds and $B_1=B^*$. Condition (2.2) is fundamental. If it is violated, then $\{X_t\}$ has an explosive character. If (2.2) holds and $B_1\neq B^*$, then the process $\{X_t\}$ tends to be stationary when $t\to\infty$.

2.2. Covariance function of a non-stationary AR process

Denote $R_{st}=\mathrm{Cov}(X_s,X_t)$ and $B_t=\mathrm{Var}(X_t,\ldots,X_{t+n-1})'$. Then

$$B_t = M^{t-1}B_1M'^{t-1} + \sum_{j=0}^{t-2} M^j\Lambda M'^j \tag{2.3}$$

and

$$\left\|\begin{matrix} R_{t,t+k} & \cdots & R_{t,t+k+n-1} \\ \cdots & \cdots & \cdots \\ R_{t+n-1,t+k} & \cdots & R_{t+n-1,t+k+n-1}\end{matrix}\right\| = B_tM'^k \tag{2.4}$$

for $t\geq 1$ and $k\geq 0$ [see Anděl (1977) and Anděl (1979)]. From (2.3) and (2.4) we shall derive explicit formulas for R_{st} in some special cases. It is also possible to get a recurrent procedure for R_{st} (see the end of Section 2.4).

2.3. Covariance function of AR(1)

If $|a_1|\neq 1$, then we have from (2.3)

$$R_{tt} = a_1^{2t-2}R_{11}+(1-a_1^2)^{-1}(a_1^{2t-2}-1)\sigma^2,$$

whereas for $|a_1|=1$ we obtain

$$R_{tt}=R_{11}+(t-1)\sigma^2.$$

In the both cases

$$R_{t,t+k} = (-a_1)^kR_{tt}, \quad t\geq 1, \quad k\geq 0.$$

2.4. Covariance function of AR(2)

Denote by z_1 and z_2 the roots of equation $z^2+a_1z+a_2=0$. Let $z_1\neq z_2$. Then from the Perron's formula we have

$$M^k = z_1^kH_1+z_2^kH_2, \quad k = 0,1,\ldots$$

where

$$H_1 = (z_1-z_2)^{-1}(M-z_2I), \quad H_2 = (z_2-z_1)^{-1}(M-z_1I).$$

Using (2.3) and (2.4) we obtain after some computations the following results.

If $z_1 \neq \pm 1$, $z_2 \neq \pm 1$, then

$$R_{tt} = \sigma^2(z_1-z_2)^{-2}\Big\{(z_1^2-1)^{-1}(z_1^{2t-2}-1)-2(z_1z_2-1)^{-1}[(z_1z_2)^{t-1}-1]+$$

$$+ (z_2^2-1)^{-1}(z_2^{2t-2}-1)\Big\}+(z_1-z_2)^{-2}[(z_1^{t-1}-z_2^{t-1})^2R_{22} -$$

$$- 2(z_1^{t-1}-z_2^{t-1})(z_1^{t-1}z_2-z_1z_2^{t-1})R_{12}+(z_1^{t-1}z_2-z_1z_2^{t-1})^2R_{11}],$$

$$R_{t,t+1} = -a_1\sigma^2(z_1-z_2)^{-2}\Big\{(z_1^2-1)^{-1}(z_1^{2t-2}-1)-$$

$$- 2(z_1z_2-1)^{-1}[(z_1z_2)^{t-1}-1]+(z_2^2-1)^{-1}(z_2^{2t-2}-1)\Big\} -$$

$$- (z_1-z_2)^{-2}[(z^{t-1}-z_2^{t-1})^2(a_2R_{12}+a_1R_{11}) +$$

$$+ (z_1^{t-1}-z_2^{t-1})(z_1^{t-1}z_2-z_1z_2^{t-1})(R_{22}-a_2R_{11}-a_1R_{12}) -$$

$$- (z_1^{t-1}z_2-z_1z_2^{t-1})^2R_{12}]-\sigma^2(z_1-z_2)^{-2}\Big\{z_2(z_1^2-1)^{-1}(z_1^{2t-2}-1)-$$

$$- (z_1+z_2)(z_1z_2-1)^{-1}[(z_1z_2)^{t-1}-1]+z_1(z_2^2-1)^{-1}(z_2^{2t-2}-1)\Big\},$$

$$R_{t,t+k} = -(z_1-z_2)^{-1}(z_1^kz_2-z_1z_2^k)R_{tt}+(z_1-z_2)^{-1}(z_1^k-z_2^k)R_{t,t+1}.$$

If $z_1=\pm 1$ and $z_2\neq\pm 1$, the above formulas hold with the exception that we must write $t-1$ instead of $(z_1^2-1)^{-1}(z_1^{2t-2}-1)$.

If $z_1=1$ and $z_2=-1$, then

$$R_{tt} = \begin{cases} \sigma^2(t-1)/2 + R_{11} & \text{for } t \text{ odd}, \\ \sigma^2(t-2)/2 + R_{22} & \text{for } t \text{ even}, \end{cases}$$

$$R_{t,t+1} = R_{12},$$

$$R_{t,t+k} = \begin{cases} R_{t,t+1} & \text{for } k \text{ odd,} \\ R_{tt} & \text{for } k \text{ even.} \end{cases}$$

If z_1 and z_2 are not real, then $z_{12}=z\exp(\pm i\varphi)$, where $z>0$ and $\varphi\in(0,\pi)$. Then for $z\neq 1$

$$R_{tt} = -\sigma^2(2z^2\sin^2\varphi)^{-1}\Big\{(1+z^4-2z^2\cos 2\varphi)^{-1}[z^{2t}\cos(2t-4)\varphi -$$
$$- z^{2t-2}\cos(2t-2)\varphi - z^2\cos 2\varphi+1] - (z^2-1)^{-1}(z^{2t-2}-1)\Big\} +$$
$$+ (z^2\sin^2\varphi)^{-1}[z^{2t-2}\sin^2(t-1)\varphi\cdot R_{22} -$$
$$- 2z^{2t-1}\sin(t-1)\varphi\cdot\sin(t-2)\varphi\cdot R_{12} + z^{2t}\sin^2(t-2)\varphi\cdot R_{11}] ,$$

$$R_{t,t+1} = \sigma^2 a_1(2z^2\sin^2\varphi)^{-1}\Big\{(1+z^4-2z^2\cos 2\varphi)^{-1}[z^{2t}\cos(2t-4)\varphi -$$
$$- z^{2t-2}\cos(2t-2)\varphi - z^2\cos 2\varphi+1] - (z^2-1)^{-1}(z^{2t-2}-1)\Big\} -$$
$$- (z^2\sin^2\varphi)^{-1}[z^{2t-2}\sin^2(t-1)\varphi\cdot(a_2R_{12}+a_1R_{22}) +$$
$$+ z^{2t-2}\sin(t-1)\varphi\cdot\sin(t-2)\varphi\cdot(R_{22}-a_2R_{11}-a_1R_{12}) -$$
$$- z^{2t}\sin^2(t-2)\varphi\cdot R_{12}] + \sigma^2(2z^2\sin^2\varphi)^{-1}\Big\{(1+z^4-2z^2\cos 2\varphi)^{-1} \times$$
$$\times [z^{2t+1}\cos(2t-5)\varphi - z^{2t-1}\cos(2t-3)\varphi - z^3\cos 3\varphi + z\cos\varphi] -$$
$$- (z^2-1)^{-1}z(z^{2t-2}-1)\cos\varphi\Big\},$$

$$R_{t,t+k} = z^{k-1}[-z\sin(k-1)\varphi\cdot R_{tt} + \sin k\varphi\cdot R_{t,t+1}]/\sin\varphi .$$

For $z=1$ we must again write $t-1$ instead of $(z^2-1)^{-1}(z^{2t-2}-1)$.

Let $z_1=z_2=z\neq 0$. Then $M^k=z^kH_1+kz^{k-1}H_2$, where $H_1=I$, $H_2=M-zI$. If $z^2\neq 1$, then

$$R_{tt} = \sigma^2 z^{-2}\Big\{2z^4(1-z^2)^{-3}(1-z^{2t-6})+z^2(1-z^2)^{-2}[1-(2t-5)z^{2t-4}] -$$

$$- (t-2)^2(1-z^2)^{-1}z^{2t-2}\Big\}+(t-1)^2 z^{2t-4}R_{22} -$$

$$- 2(t-1)(t-2)z^{2t-3}R_{12}+(t-2)^2 z^{2t-2}R_{11},$$

$$R_{t,t+1} = \sigma^2 z^{-1}\Big\{2z^4(1-z^2)^{-3}(1-z^{2t-6})+z^2(1-z^2)^{-2}[1-(2t-5)z^{2t-4}] -$$

$$- (t-2)^2 z^{2t-2}(1-z^2)^{-1}+z^2(1-z^2)^{-2}(1-z^{2t-4})-(t-2)z^{2t-2}(1-z^2)^{-1}\Big\} -$$

$$- (t-1)^2 z^{2t-4}(a_2R_{12}+a_1R_{22})-(t-1)(t-2)z^{2t-3}(R_{22}-a_2R_{11}-a_1R_{12}) +$$

$$+ (t-2)^2 z^{2t-2}R_{12} \quad ,$$

$$R_{t,t+k} = -(k-1)z^k R_{tt} + kz^{k-1}R_{t,t+1} \; .$$

If $z_1 = z_2 = 1$, then

$$R_{tt} = (\sigma^2/6)(t-1)(t-2)(2t-3) +(t-1)^2R_{22}-2(t-1)(t-2)R_{12}+(t-2)^2R_{11},$$

$$R_{t,t+1} = (\sigma^2/3)t(t-1)(t-2)-(t-1)^2(R_{12}-2R_{22}) -$$

$$-(t-1)(t-2)(R_{22}-R_{11}+2R_{12})+(t-2)^2R_{12},$$

$$R_{t,t+k} = -(k-1)R_{tt}+k\,R_{t,t+1} \; .$$

If $z_1 = z_2 = -1$, we have

$$R_{tt} = (\sigma^2/6)(t-1)(t-2)(2t-3)+(t-1)^2R_{22}+2(t-1)(t-2)R_{12}+(t-2)^2R_{11},$$

$$R_{t,t+1} = -(\sigma^2/3)t(t-1)(t-2)-(t-1)^2(R_{12}+2R_{22}) +$$

$$+ (t-1)(t-2)(R_{22}-R_{11}-2R_{12})+(t-2)^2R_{12},$$

$$R_{t,t+k} = (-1)^{k+1}[(k-1)R_{tt}+k\,R_{t,t+1}].$$

If t and k are not too large, we can use in all the cases the recurrent procedure

$$\begin{Vmatrix} R_{t+1,t+1} & R_{t+1,t+2} \\ R_{t+1,t+2} & R_{t+2,t+2} \end{Vmatrix} = M \begin{Vmatrix} R_{tt} & R_{t,t+1} \\ R_{t,t+1} & R_{t+1,t+1} \end{Vmatrix} M' + \Lambda$$

and

$$\begin{Vmatrix} R_{t,t+k+1} & R_{t,t+k+2} \\ R_{t+1,t+k+1} & R_{t+1,t+k+2} \end{Vmatrix} = \begin{Vmatrix} R_{t,t+k} & R_{t,t+k+1} \\ R_{t+1,t+k} & R_{t+1,t+k+1} \end{Vmatrix} M' ,$$

which follows from (2.3) and from (2.4).

The formulas for the covariance function for stationary AR(1) and AR(2) are well known and therefore they will not be repeated here.

3. Forward extrapolation

Consider model (2.1). Let $X_1,\dots,X_{m-1}$ be known, where $m \ge n+1$. Then the best linear (forward) extrapolation $\hat{X}_m$ of X_m is

$$\hat{X}_m = -a_1 X_{m-1} - \dots - a_n X_{m-n}$$

and the residual variance is

$$\delta^2 = E(X_m - \hat{X}_m)^2 = \sigma^2 .$$

The extrapolation of $X_{m+1}, X_{m+2}, \dots$ is also quite similar to the procedure for stationary AR processes.

4. Backward extrapolation

4.1. Introduction

Consider again model (2.1). Let $X_{m+1}, X_{m+2}, \dots$ be known. Denote by $\tilde{X}_m$ the best linear estimator of X_m. Recall that $B_{m+1} = \mathrm{Var}(X_{m+1},\dots,X_{m+n})'$ and put $E_{m+1} = B_{m+1}^{-1}$. If B_1 is regular, then $B_2, B_3, \ldots$ are also regular; see Anděl (1971), Theorem 6. The formula for $\tilde{X}_m$ is

$$\tilde{X}_m = (R_{m,m+1},\ldots,R_{m,m+n})E_{m+1}(X_{m+1},\ldots,X_{m+n})'$$

[see Anděl (1977)]. This is the result for general non-stationary AR models. If $\{X_t\}$ is stationary, then it can be proved that the backward extrapolation is given by the same coefficients as the forward extrapolation, e.g.

$$\tilde{X}_m = -a_1X_{m+1} - \ldots - a_nX_{m+n} \ .$$

4.2. Backward extrapolation in AR(1)

If n=1, then we have

$$\tilde{X}_m = -a_1(R_{mm}/R_{m+1,m+1})X_{m+1}$$

and

$$\omega_m^2 = E(X_m-\tilde{X}_m)^2 = R_{mm}-a_1^2R_{mm}^2/R_{m+1,m+1} \ .$$

4.3. Backward extrapolation in AR(2)

Denote

$$\Delta_m = R_{m+1,m+1}R_{mm}-R_{m,m+1}^2 \ ,$$

$$\alpha = (a_2^2\Delta_m+\sigma^2R_{m+1,m+1})^{-1}(-a_1a_2\Delta_m+\sigma^2R_{m,m+1}),$$

$$\beta = -(a_2^2\Delta_m+\sigma^2R_{m+1,m+1})^{-1}a_2\Delta_m \ .$$

Then

$$\tilde{X}_m = \alpha X_{m+1}+\beta X_{m+2}$$

and

$$\omega_m^2 = E(X_m-\tilde{X}_m)^2 = R_{mm}+\alpha^2 R_{m+1,m+1}+\beta^2 R_{m+2,m+2} -$$

$$- 2\alpha R_{m,m+1}-2\beta R_{m,m+2}+2\alpha\beta R_{m+1,m+2} .$$

5. Interpolation

Anděl (1979) proved that in non-stationary AR(n) models the best linear interpolation depends only on n previous and n subsequent known variables X_t. Till the end of this paper we shall assume that model (2.1) holds and that $X_1,\dots,X_{m-1}$, $X_{m+1},X_{m+2},\dots$ are known, where $m-1\geq n$. Then the best linear interpolation X_m^o of the variable X_m is given by

$$X_m^o = -\left(\sum_{i=0}^{n} a_i^2\right)^{-1}\left[\sum_{k=1}^{n}\left(\sum_{j=0}^{n-k} a_{j+k}a_j\right)(X_{m+k}+X_{m-k})\right].$$

After a calculation we get the corresponding residual variance

$$\tau^2 = E(X_m-X_m^o)^2 = \sigma^2\left(\sum_{i=0}^{n} a_i^2\right)^{-1}.$$

6. Damsleth's method

6.1. Stationary case

Let the process $\{X_t\}$ be stationary and let R_t be its covariance function. Then we obtain

$$X_m^* = -\left(R_o-\sigma^2+\sum_{j=1}^{n}\sum_{k=1}^{n} a_j a_k R_{j+k}\right)^{-1}\left(\sum_{k=1}^{n} a_k R_k\right)(\tilde{X}_m+\hat{X}_m)$$

and

$$\lambda^2 = E(X_m-X_m^*)^2 = R_o-2\left(R_o-\sigma^2+\sum_{j=1}^{n}\sum_{k=1}^{n} a_j a_k R_{j+k}\right)^{-1}\left(\sum_{k=1}^{n} a_k R_k\right)^2.$$

6.2. Non-stationary AR(1)

In this case

$$\hat{X}_m = - a_1 X_{m-1}, \quad \tilde{X}_m = - a_1 (R_{mm}/R_{m+1,m+1}) X_{m+1} .$$

In Section 5 we made a remark that in AR(n) models the interpolation X_m^o depends only on n previous and n subsequent observations. In our case it means that X_m^o is a linear function of X_{m-1} and X_{m+1}. From here we can see that the Damleth's method leads to estimator X_m^* which is equal to X_m^o. This result can be checked also by direct calculation. All these methods give

$$X_m^* = - a_1 (1+a_1^2)^{-1} (X_{m-1}+X_{m+1}),$$

$$\lambda^2 = E(X_m - X_m^*)^2 = \sigma^2 (1+a_1^2)^{-1}.$$

6.3. Non-stationary AR(2)

Here the variables X_m^* and X_m^o will be different in general. From Sections 3 and 4.3 we remember that

$$\hat{X}_m = -a_1 X_{m-1} - a_2 X_{m-2}, \quad \tilde{X}_m = \alpha X_{m+1} + \beta X_{m+2}.$$

Denote

$$R = \mathrm{Cov}(X_m, \hat{X}_m), \; S = \mathrm{Cov}(X_m, \tilde{X}_m), \; P = \mathrm{Var}\, \hat{X}_m, \; Q = \mathrm{Var}\, \tilde{X}_m, \; L = \mathrm{Cov}(\hat{X}_m, \tilde{X}_m).$$

Then

$$R = -a_1 R_{m-1,m} - a_2 R_{m-2,m} , \qquad S = \alpha R_{m,m+1} + \beta R_{m,m+2} ,$$

$$P = a_1^2 R_{m-1,m-1} + 2a_1 a_2 R_{m-2,m-1} + a_2^2 R_{m-2,m-2} ,$$

$$Q = \alpha^2 R_{m+1,m+1} + 2\alpha\beta R_{m+1,m+2} + \beta^2 R_{m+2,m+2} ,$$

$$L = -a_1 \alpha R_{m-1,m+1} - a_2 \alpha R_{m-2,m+1} - a_1 \beta R_{m-1,m+2} - a_2 \beta R_{m-2,m+2}.$$

After a calculation we obtain that

$$X_m^* = u\hat{X}_m + v\tilde{X}_m ,$$

where

$$u = (PQ-L^2)^{-1}(RQ-SL), \quad v = (PQ-L^2)^{-1}(PS-RL),$$

and

$$\lambda_m^2 = E(X_m-X_m^*)^2 = R_{mm} - (PQ-L^2)^{-1}(PS^2-2RSL+QR^2).$$

7. Numerical comparisons

Some numerical results for stationary AR(2) processes are given in Tab. 7.1. Without loss of generality we assumed $\sigma^2=1$. The rows of the table are ordered according to increasing values of R_o.

Tab. 7.1

z_1	z_2	a_1	a_2	R_o	$\delta^2=\omega^2$	λ^2	τ^2
0.2	0.1	-0.3	0.02	1.095	1	0.917	0.917
0.9i	-0.9i	0	0.81	2.908	1	0.604	0.604
-0.95	0.8	0.15	-0.76	3.885	1	0.654	0.625
0.95	-0.9	-0.05	-0.855	4.220	1	0.585	0.577
0.95	-0.6	-0.35	-0.57	4.389	1	0.720	0.691
0.95	0.1	-1.05	0.095	12.535	1	0.474	0.474
0.99	-0.9	-0.09	-0.891	15.245	1	0.584	0.555
$0.98\exp(\pm i\pi/6)$		-1.697	0.960	51.462	1	0.476	0.208
$0.98\exp(\pm 5i\pi/6)$		1.697	0.960	51.462	1	0.476	0.208
0.95	0.9	-1.85	0.855	690.586	1	0.386	0.194
-0.95	-0.9	1.85	0.855	690.586	1	0.386	0.194

One can see that really with growing R_0 the value of λ^2 is considerably smaller than 1, which is the value of δ^2 as well as of ω^2. However, τ^2 tends also to be smaller than λ^2. We can conclude that when the roots z_1 and z_2 approach to the boundary of the unit circle, the Damsleth's method is much better than the forward extrapolation (or the backward extrapolation), but just in these cases the best linear interpolation reduces the residual variance also substantially in comparison with the Damsleth's method.

Some results were obtained also for non-stationary AR(2) processes. In this case it was necessary to choose values of R_{11}, R_{12}, R_{22} and m. We took

$$R_{11}=2,\ R_{12}=1,\ R_{22}=3,\ m=10.$$

Table 7.2 summarizes the results.

Tab. 7.2

z_1	z_2	a_1	a_2	$R_{10,10}$	δ^2	ω^2_{10}	λ^2_{10}	τ^2_{10}
1	-1	0	-1	7	1	0.873	0.500	0.500
$1.02\exp(\pm i\pi/2)$		0	1.040	8.641	1	0.832	0.481	0.480
$1.05\exp(\pm i\pi/2)$		0	1.102	12.037	1	0.768	0.452	0.451
$1.1\exp(\pm i\pi/2)$		0	1.210	21.531	1	0.661	0.406	0.406
$1.02\exp(\pm i\pi/6)$		-1.767	1.040	33.570	1	0.791	0.398	0.192
$1.05\exp(\pm i\pi/6)$		-1.819	1.105	47.158	1	0.736	0.392	0.181
$1.1\exp(\pm i\pi/6)$		-1.905	1.210	85.956	1	0.642	0.373	0.164
1	1	-2	1	431	1	0.701	0.304	0.166
-1	-1	2	1	719	1	0.712	0.309	0.167
$1.5\exp(\pm i\pi/2)$		0	2.250	2131.960	1	0.198	0.165	0.165
$1.5\exp(\pm i\pi/6)$		-2.598	2.250	11016.686	1	0.197	0.164	0.078

In these non-stationary cases the backward extrapolation has a smaller residual variance than the forward extrapolation. In most cases λ_{10}^2 is much smaller than ω_{10}^2, so that the Damsleth's method is substantially better than the one-sided extrapolation. However, also here in many cases the best linear interpolation gives substantially better results than the Damsleth's method.

REFERENCES

ANDĚL J. (1971). On the multiple autoregressive series. Ann. Math. Statist. 42, 755-759.

ANDĚL J. (1976). Statistical Analysis of Time Series (in Czech). SNTL, Prague.

ANDĚL J. (1977). On the backward extrapolation of non-stationary autoregressive series. Trans. 7th Prague Conf. Inf. Th. Vol. A, 29-36.

ANDĚL J. (1979). On interpolation of multiple autoregressive processes. Contributions to Statistics (Jaroslav Hájek Memorial Volume) 13-17. Academia, Prague.

BRUBACHER S. R. and WILSON T. G. (1976). Interpolating time series with application to the estimation of holiday effects on electricity demand. Appl. Statist. 25, 107-116.

DAMSLETH E. (1980). Interpolating missing values in a time series. Scand. J. Statist. 7, 33-39.

Jiří Anděl
Matematicko-Fyzikální Fakulta
Univerzity Karlovy
Sokolovská 83
CS-18600 Praha 8

Proc. of the 4th Pannonian Symp. on Math. Stat.. Bad Tatzmannsdorf, Austria 1983
W. Grossmann, G. Pflug, I. Vincze, W. Wertz, eds

A LIMIT THEOREM FOR EHRENFEST URN MODEL

G. BARÓTI

Budapest

Consider two urns A, B and N balls numbered 1,2,...,N. Initially, let each ball be in urn A. Choose randomly an integer from 1,2,...,N and find the ball with that number on it and transfer to the other urn. Let $X_{n,N}$ denote the number of balls in urn A after the n-th transfer. This model is essentially known as the Ehrenfest urn model. At the third PSMS (see [1]) we proved that the limit distribution of the standardized $X_{n,N}$ is normal, whenever $n,N\to\infty$, with $n^2N^{-1}\to\infty$. Now we shall prove the following local limit theorem:

THEOREM. If n and N-k have the same parity, $n,N\to\infty$, with $n^2N^{-1}\to\infty$, then

$$P(X_{n,N}=k) = \frac{2}{\sigma_{n,N}\sqrt{2\pi}} e^{-\frac{(k-m_{n,N})^2}{2\sigma^2_{n,N}}}$$

uniformly relative to $\frac{k-m_{n,N}}{\sigma_{n,N}}$ in every finite interval, where $m_{n,N} = E(X_{n,N})$ and $\sigma^2_{n,N} = \mathrm{var}(X_{n,N})$.

PROOF. We suppose that n balls are randomly assigned to N urns in such a way that each urn has the same probability N^{-1} of assignment. Denoting the number of urns containing an even number of balls by $\tilde{X}_{n,N}$, it is easy to see that for fixed N the sequences $\{X_{n,N}\}$ and $\{\tilde{X}_{n,N}\}$ are Markov chains with common state space and transition matrix. Thus

$$P(X_{n,N} = k) = P(\tilde{X}_{n,N} = k). \tag{1}$$

With the aid of a simple combinatorial consideration we have

$$N^n P(\tilde{X}_{n,N}=k) = \binom{N}{k} \sum_{I} \left(\prod_{j=1}^{k} \binom{n-2s_{j-1}}{2i_j} \right) \prod_{\ell=k+1}^{N} \binom{n-2s_{\ell-1}-\ell+k+1}{2i_\ell+1} \tag{2}$$

where $s_m=i_1+i_2+\ldots+i_m$ $(s_0=0)$ and $I=\{(i_1,i_2,\ldots,i_N) \mid 2i_1+2i_2+\ldots+ +\ldots+2i_k+2i_{k+1}+1+\ldots+2i_N+1\}$. Indeed, if k urns contain an even number of balls, then there are $\binom{N}{k}$ ways to choose these urns from N urns. If the jth one of these urns contains $2i_j$ balls $(j=1,2,\ldots,k)$ and the ℓth one of the remaining N-k urns contains $2i_\ell+1$, then there are

$$\binom{n}{2i_1}\binom{n-2i_1}{2i_2}\cdots\binom{n-2s_{k-1}}{2i_k}\binom{n-2s_k}{2i_{k+1}+1}\cdots\binom{n-2s_k-(2i_{k+1}+1)-\ldots-(2i_{N-1}+1)}{2i_N+1}$$

ways to assign n balls.

Let $Y_1^{(1)}, Y_2^{(1)}, \ldots, Y_1^{(2)}, Y_2^{(2)}, \ldots$ be independent random variables with the following distributions:

$$P(Y_i^{(1)} = 2\ell) = \frac{\alpha^{2\ell}}{(2\ell)!\,\mathrm{ch}\,\alpha}, \tag{3}$$

$$P(Y_i^{(2)} = 2\ell+1) = \frac{\alpha^{2\ell+1}}{(2\ell+1)!\,\mathrm{sh}\,\alpha} \quad (i=1,2,\ldots;\ \ell=0,1,2,\ldots) \tag{4}$$

where $\alpha=nN^{-1}$.

Set

$$S_{k,N} = Y_1^{(1)} + Y_2^{(1)} + \ldots + Y_k^{(1)} + Y_{k+1}^{(2)} + \ldots + Y_N^{(2)}$$

and

$$p_\alpha = e^{-\alpha}\mathrm{ch}\alpha .$$

Now (2) can be written as

$$P(\tilde{X}_{n,N}=k) = \frac{n!e^n}{n^n}\binom{N}{k} p_\alpha^k (1-p_\alpha)^{N-k} P(S_{k,N}=n) \tag{5}$$

Using Stirling's formula and the normal approximation of binomial distribution we obtain

$$\frac{n!e^n}{n^n}\binom{N}{k} p_\alpha^k (1-p_\alpha)^{N-k} = \sqrt{\frac{\alpha}{p_\alpha(1-p_\alpha)}}\, e^{-\frac{(k-Np_\alpha)^2}{2Np_\alpha(1-p_\alpha)}} (1+\Theta(1)) \tag{6}$$

uniformly relative to $\frac{k-Np_\alpha}{\sqrt{Np_\alpha(1-p_\alpha)}}$ in every finite interval, since under the assumptions of our Theorem $n\to\infty$ and $Np_\alpha(1-p_\alpha)\to\infty$. The approximation of $P(S_{k,N}=n)$ is given by the following lemma.

LEMMA. Under the conditions of our Theorem

$$P(S_{k,N}=n) = \frac{2}{\sigma\sqrt{2\pi}} e^{-\frac{(n-m)^2}{2\sigma^2}} (1+\Theta(1)) \tag{7}$$

uniformly relative to $\frac{n-m}{\sigma}$ in every finite interval, where $m=E(S_{k,N})$ and $\sigma^2=\mathrm{var}(S_{k,N})$.

Proof of the Lemma. We shall follow the method of the classical proof of the local limit theorem for independent, identically distributed and integer valued random variables (see [3]).

Let $\varphi(t)$ be the characteristic function of $S_{k,N}$,

$$\Psi(t) = e^{-\frac{imt}{\sigma}} \varphi\left(\frac{t}{\sigma}\right) \text{ and } u = \frac{n-m}{\sigma} .$$

By the inversion formula we obtain

$$P(S_{k,N}=n) = \frac{1}{\pi}\int_{-\pi/2}^{\pi/2} e^{-int} \varphi(t)dt$$

supposing that n and N-k have the same perity. The difference

$$R_N = \pi(\sigma P(S_{k,N}=n) - \frac{2}{\sqrt{2\pi}} e^{-\frac{u^2}{2}})$$

can be written as the sum of four integrals

$$R_N = I_1 + I_2 + I_3 + I_4$$

where

$$I_1 = \int_{|t|\le C} e^{-iut}(\Psi(t)-e^{-\frac{t^2}{2}})dt,$$

$$I_2 = \int_{C\le|t|\le\varepsilon\sigma} e^{-iut}\,\Psi(t)dt,$$

$$I_3 = \int_{\varepsilon\sigma\le|t|\le\frac{\pi\sigma}{2}} e^{-iut}\,\Psi(t)dt,$$

$$I_4 = -\int_{C<|t|} e^{-iut-\frac{t^2}{2}} dt$$

C and ε are arbitrary positive constants).

The characteristic function of $Y_1^{(1)}$ resp. $Y_1^{(2)}$ is

$$\varphi_1(t) = \frac{ch(\alpha e^{it})}{ch\alpha} \quad \text{resp.} \quad \varphi_2(t) = \frac{sh(\alpha e^{it})}{sh\alpha}. \tag{8}$$

Introduce the following notations

$$E(Y_1^{(j)})=m_j,\ var(Y_1^{(j)})=\sigma_j,\ f_j(t)=\log(e^{-im_jt}\varphi_j(t))\ (j=1,2) \tag{9}$$

$$h_1(x) = thx,\ h_2(x) = cthx. \tag{10}$$

By simple calculation we get

$$m_j = \alpha h_j(\alpha),\ \sigma_j^2 = \alpha^2+\alpha h_j(\alpha) - \alpha^2 h_j(x), \tag{11}$$

$$f_j'''(t) = -izh_j(z)-3iz^2(1-h_j^2(z))+2iz^3h_j(z)(1-h_j^2(z)) \quad (z=\alpha e^{it}) \tag{12}$$

$$\sigma_j^2 \sim \alpha \quad (\alpha\to\infty)\ (j=1,2),\ \sigma_1^2\sim 2\alpha^2,\ \sigma_2^2 \sim \frac{2}{3}\alpha^2 \qquad (\alpha\to 0). \qquad (13)-(14)$$

One can readily verify that $|\sigma_j^{-2}\ f_j'''(t)|$ remains bounded in a neighbourhood of $t=0$ while α varies $(j=1,2)$.

(If $\alpha\to\infty$, use the following relations

$$1-\mathrm{th}^2 z=\mathrm{ch}^{-2} z,\ \mathrm{cth}^2 z-1=\mathrm{sh}^{-2} z,$$

$$|\mathrm{th}(\alpha e^{it})|\le\mathrm{cth}(\alpha\cos t),\ |\mathrm{cth}(\alpha e^{it})|\le\mathrm{cth}(\alpha\cos t),$$

$$|\mathrm{ch}(\alpha e^{it})|\ge\mathrm{sh}(\alpha\cos t),\ |\mathrm{sh}(\alpha e^{it})|\ge\mathrm{sh}(\alpha\cos t)$$

and if $\alpha\to 0$, use Taylor series expansions to obtain

$$f_1'''(t)=-4iz^2+\Theta(z^3),\ f_2'''(t)=-\frac{4}{3}iz^2+\Theta(z^3),$$

finally apply the asymptotic relations (13) - (14)).

Thus

$$f_j(t) = \frac{\sigma_j^2 t^2}{2}(1+\Theta(t)) \qquad (t\to 0,\ j=1,2)\ . \tag{15}$$

Since

$$m = km_1 + (N-k)m_2$$

$$\sigma^2= k\sigma_1^2 + (N-k)\ \sigma_2^2$$

$$\Psi(t) = e^{kf_1(\frac{t}{\sigma}) +(N-k)f_2(\frac{t}{\sigma})}$$

therefore from (13)-(15) it follows that

$$\Psi(t) \to e^{-\frac{t^2}{2}}$$

uniformly in every finite t-interval, thus

$$I_1 \to 0. \tag{16}$$

On the other hand, by virtue of relation (15), there exist positive constants c_1 and ε, such that for $0 \le |t| < \varepsilon$

$$|\varphi_j(t)| \le e^{-c_1 \sigma_j^2 t^2} \qquad (j=1,2)$$

hence

$$|I_2| \le \int_{C \le |t|} e^{-c_1 t^2} dt. \tag{17}$$

If $\varepsilon \le |t| \le \frac{\pi}{2}$, we shall distinguish three cases according as α is sufficiently large or α is bounded from above and below too or α is sufficiently small.

a) $\alpha \ge \alpha_2$.

$$|\varphi_1(t)| \le \frac{\operatorname{ch}(\alpha\cos t)}{\operatorname{ch}\alpha} = e^{-\alpha\sin^2\frac{t}{2}} \frac{e^{-\alpha\sin^2\frac{t}{2}} + e^{-\frac{\alpha}{2}(1+3\cos t)}}{1+e^{-2\alpha}} \le e^{-\alpha\sin^2\frac{t}{2}}$$

and in a similar manner

$$|\varphi_2(t)| \le \frac{\operatorname{ch}(\alpha\cos t)}{\operatorname{sh}\alpha} \le e^{-\alpha\sin^2\frac{t}{2}}$$

b) $0 < \alpha_1 \le \alpha \le \alpha_2$.

It follows from the definition of $\varphi_j(t)$ that $|\varphi_j(t)|$ is continuous and less than 1 on the rectangle

$\{(t,\alpha) \mid \varepsilon \le |t| \le \frac{\pi}{2},\ \alpha_1 \le \alpha \le \alpha_2\}$.

Thus

$$|\varphi_j(t)| \le q < 1 \qquad (j=1,2),$$

where q is a constant.

c) $\alpha < \alpha_1$.

By Taylor series expansions we obtain

$$|\varphi_1(t)|^2 = \frac{ch^2(\alpha\cos t) - \sin^2(\alpha\sin t)}{ch^2\alpha} = 1-2\alpha^2 \sin^2 t + \Theta(\alpha^3)$$

$$|\varphi_2(t)|^2 = \frac{ch^2(\alpha\cos t) - \cos^2(\alpha\sin t)}{sh^2\alpha} = 1-\frac{2}{3}\alpha^2 \sin t + \Theta(\alpha^3)$$

We can summarize these results as follows. If $\varepsilon \le |t| \le \frac{\pi}{2}$, then there exists a positive constant c_2 such that $|\varphi_j(t)| \le e^{-c_2\alpha}$ ($j=1,2$) whatever α be.

Therefore

$$|I_3| \le \sigma e^{-c_2 N\alpha} \to 0. \tag{18}$$

(Note that, if $\alpha\to\infty$, then $\sigma^2=\Theta(N\alpha)$; if $\alpha_1\le\alpha\le\alpha_2$, then $\sigma^2=\Theta(N)$ and if $\alpha\to 0$, then $\sigma^2=\Theta(N\alpha^2)$).

Finally it is trivial that

$$|I_4| \le \int_{C\le|t|} e^{-\frac{t^2}{2}} dt. \tag{19}$$

Consequently, if C is sufficiently large then $|I_2|$ and $|I_4|$ can be arbitrary small (see (17) and (19)) thus it follows from (16)-(19) that $R_N\to 0$. This completes the proof of the Lemma.

Now let us return to the proof of the Theorem. For the expectation and variance of $X_{n,N}$ in [1] we have obtained the following formulas.

$$E(X_{n,N}) = \frac{N}{2}\left(1+\left(1-\frac{2}{N}\right)^n\right),$$

$$\operatorname{var}(X_{n,N}) = \frac{N}{4}\left(1+(N-1)\left(1-\frac{4}{N}\right)^n - N\left(1-\frac{2}{N}\right)^{2n}\right).$$

Supposing that the conditions of our Theorem are met that is $n,N\to\infty$ with $N\alpha^2\to\infty$, we have

$$E(X_{n,N}) \sim \frac{N}{2}(1 + e^{-2\alpha}),$$

$$\mathrm{var}(X_{n,N}) \sim \frac{N}{4}(1 - e^{-4\alpha}-4\alpha e^{-4\alpha}) .$$

Setting

$$x = \frac{k - \frac{N}{2}(1 + e^{-2\alpha})}{\sqrt{\frac{N}{4}(1 - e^{-4\alpha}-4\alpha e^{-4\alpha})}} ,$$

by simple calculations we obtain

$$\sigma^2 = \frac{N\alpha(1-e^{-4\alpha}-4\alpha e^{-4\alpha})}{1-e^{-4\alpha}} (1+\Theta(1))$$

and

$$\frac{(n-m)^2}{2\sigma^2} = \frac{2\alpha e^{-4\alpha}x^2}{1-e^{-4\alpha}} (1+\Theta(1)) = \Theta(\alpha^2).$$

Hence by the Lemma

$$P(S_{k,N}=n) = \frac{2}{\sigma\sqrt{2\pi}} e^{-\frac{2\alpha e^{-4\alpha}x^2}{1-e^{-4\alpha}}} (1+\Theta(1)) \qquad (20)$$

uniformly in every finite x-interval.

Since

$$\frac{(k-Np_\alpha)^2}{Np_\alpha(1-p_\alpha)} = \Theta(\alpha^2)$$

the statement of the Theorem follows from (1), (5) and (20).

Remark. In the classical occupancy problem (that is N urns and n balls are considered and the balls are assigned randomly in the urns) let us denote by $X_{n,N}^{(r)}$ the number of urns containing exactly r balls. In [2] many local limit theorems are proved for $X_{n,N}^{(r)}$. Our proof follows the idea of those theorems.

REFERENCES

BARÓTI, G. Limit distribution for Ehrenfest's urn model. Proceedings of 3rd Pannonian Symposion on Math. Stat., Visegrád, 1982. (edited by J. Mogyoródi).

KOLCHIN, V.F., SEVASTIANOV, B.A., CHISTIAKOV, V.P. Random occupancies (in Russian). Mir, Moscow, 1976.

GNEDENKO, B.V., KOLMOGOROV, A.N. Limit distribution for sums of independent random variables. Addison-Wesley Publ. Cambridge, 1954.

György Baróti
Dept. of Probability Theory
Loránd Eötvös University, Budapest,
Muzeum krt. 6-8.
H-1088. HUNGARY

Proc. of the 4th Pannonian Symp. on Math. Stat., Bad Tatzmannsdorf, Austria 1983
W. Grossmann, G. Pflug, I. Vincze, W. Wertz, eds

A NONPARAMETRIC SMALL SAMPLE THEORY OF ESTIMATION OF REGRESSION FUNCTIONS

OLAF BUNKE

Berlin

Summary

A small sample theory of estimation of regression functions is developed in a nonparametric approach. The regression function is assumed to be an element of some set of a linear space with inner product. The risk is defined by the corresponding squared norm. It is shown that every linear operator interpolating the n observations is a minimax estimator. The description of smoothness properties of the regression function by appropriate prior distributions or by ellipsoidal restrictions is discussed. The structure of corresponding Bayes and minimax procedures, which depend on the degree of smoothness, is derived. Finally the risk properties of kernel estimators are discussed, showing that they are inadmissible for special risks.

1. INTRODUCTION

A frequent problem in applications is the investigation of the dependence of a variable y (called dependent variable or predictor) on some other variables, which are called independent variables or regressors and may be collected into a vector x. It is asked for a function f(x), which describes the dependence of y on x. Usually one has n joint observations $(x_1,y_1),\dots,(x_n,y_n)$ of both variables x and y. If the observations are from a designed experiment, the values $x_1,\dots,x_n$ may be assumed to be fixed, while the observations $y_1,\dots,y_n$ are usually assumed to be realizations of random variables with expectations $f(x_1),\dots,f(x_n)$. Here f is the so called regression or response function which gives the expectation $f(x)$ of the dependent variable y for a fixed value x of the independent variable. The assumption of fixed values $x_1,\ldots,x_n$ may also be justified under a conditional inference approach (see Cox and Hinkley (1974)), even if this values are not controlled and possibly realizations of random variables.

The classical approach to the problem is parametrical and consists in fitting a parametric function $g(x,\beta)$ to the observations by least squares. Almost the whole literature on regression analysis (for surveys see e.g. the books by Seber (1977), Montgomery and Peck (1982) or Humak (1977)) is devoted to this approach and presents optimal estimators of the regression function, assuming that it has just the form $g(x,\beta)$ with some (unknown) value of the parameter β. Only few investigations dispence with this assumption and try to find estimators of the form $\hat{f}(x)=g(x,\hat{\beta})$, which have some optimality properties, as the papers by Karson, Manson and Hader (1969), Petersen (1973), Läuter (1975) and Bunke (1973) (for a survey see Humak (1977)).

A more modern approach is the estimation of the regression function by kernel or nearest neighbour estimators and other nonparametric procedures. Most of the results on optimal non-parametric estimators are asymptotic. A survey is given by Collomb (1980).

Parametric estimators in situations with an infinite number of observations, say, the realization of a stochastic process is observed, are treated in the survey paper of Thrum (1979).

As a systematic nonasymptotic theory of regression function estimation does not exist, this paper tries to answer first standard questions in a theory, which uses as risk the mean squared norm of the estimation error, namely about uniformly best, minimax and Bayes estimators. It is natural in such a problem to include smoothness assumptions on the unknown regression function. One of the main difficulties is the fact, that the regression function is not identifiable by only n observations, that is, there are different regression functions leading to the same probability distribution of the observations. A series of questions arise, if in this situation the regression function is replaced by an identifiable substitute, which is not very different from it.

We will show in Theorem 1 that interpolating the observations by a polynomial or a spline gives a minimax estimator. We discuss the description of smoothness properties by prior distributions or by ellipsoidal restrictions on the values of the regression function. Theorem 2 and 3 give results for Bayes- and minimax estimators, the last theorem for a modified risk. The structure of these estimators gives an idea for the construction of adaptive estimators, as discussed in section 2 devoted to these theorems. Section 3 is devoted to the discussion of risk properties of kernel estimators, showing that they are inadmissible, if the norm in the space of regression functions depends only on the restriction to the set of observed values $x_1,\ldots,x_n$, and that a combination of a parametric estimator and a kernel estimator may be better under realistic assumptions (theorem 4).

2. The model and optimal estimators

We assume a nonparametric model

$$E\ y = f^{\xi},\ \mathrm{Cov}\ y = \Lambda \tag{1}$$

for the vector $y = (y_1,\dots,y_n)^T$ of observations, where f is an unknown element of a set F_* of functions $f: \mathfrak{X} \to R^1$, $(\mathfrak{X} \subset R^k)$,

$$f^{\xi} = (f(x_1),\dots,f(x_n))^T \tag{2}$$

denotes the vector of values of f at the observed inputs $x_i \in \mathfrak{X}$ (with $x_i \neq x_j$ if $i \neq j$) and Λ is a known positive definite matrix. F_* is assumed to be a subset of a linear space F of functions, in which a semi-inner product $(\cdot,\cdot)$ is defined. If P denotes the probability distribution of $y-f^{\xi}$, we may characterize the distribution of y by the pair $\theta = (f,P)$. We assume that the unknown distribution P is an element of a class $\mathfrak{P}$ of distributions with zero expectation and covariance matrix Λ. We are interested in the estimation of the regression function f, defining estimators as (measurable) applications $\tilde{f}: R^n \to F$. The risk function is defined by

$$r(\theta,f) = E_{\theta}||\tilde{f}-f||^2 \tag{3}$$

using the seminorm given by $||f||^2 = (f,f)$, e.g.

$$r(\theta,f) = E\int|\tilde{f}(x) - f(x)|^2 \quad dW(x) \tag{4}$$

with some weight function W.

A first question arising in this context is: which is the analogous estimator to the GLSE and the MLE in a linear model? The answer is, that interpolating the observations by a polynomial or a spline yields a MLE under a normal distribution P, and has a minimax property like the GLSE. We need the following definition:

<u>Definition:</u> A linear operator $T: R^n \to F$ is called a linear interpolating operator, if for each $y \in R^n$ the image Ty is an inter-

polating function $f_y \in F$ interpolating at the points $(x_i, y_i): f_y(x_i) = y_i$ $(i=1,\ldots,n)$.
Such an operator may always be described in the form

$$f_y(x) = y^T h(x), \tag{5}$$

with $h=(h_1,\ldots,h_n)^T$ and $h_i \in F$ are certain functions leading to the identity matrix $I=((h_j(x_i)))$.

Theorem 1: Each linear interpolating operator $\hat{f}$ is a minimax estimator of f, if $f+T_y \in F_*$ for all $f \in F_*, y \in R^n$ and $N(0,\Lambda) \in \mathcal{P}$:

$$\sup_{f,P} r(\theta,\hat{f}) = \min_{\tilde{f}} \sup_{f,P} r(\theta,\tilde{f}) \tag{6}$$

(This and the following theorems are proved in the appendix).
This result may be considered as an extension of the wellknown minimaxity of the BLUE of a linear parameter γ in a linear model (see Humak (1977)). Its proof consists in a reduction of the problem of estimating f to the estimation of the mean f^ξ.
It is obvious, that under a normal distribution $P = N(0,\Lambda)$ each estimator $\tilde{f}$ with $\tilde{f}^\xi = y$ is a maximum likelihood estimator, y being the MLE of f^ξ.
Often it is known, that the regression function is smooth in some not welldefined sense and it is interesting, if there is a possibility of exploiting such a knowledge. For this we seek a description of smoothness, which is simple and allows the derivation of optimal estimators. Some of the many possibilities for such a description are the following:
A. Usual smoothness conditions from analysis are given by bounds for the derivative or for some integrated square norm of it:

$$|f'(x)|<c, \quad ||f'||^2<c \tag{7}$$

Since only f^ξ is identifiable, we propose in place of (7) a condition

$$\sum_{i,j:||x_i-x_j||<\varepsilon} \left(\frac{|f(x_i)-f(x_j)|}{||x_i-x_j||} \right)^2 < c \tag{8}$$

with some small positive ε. (8) is equivalent to $||f^\xi||_M^2 < c$ with $||y||_M^2 = y^T My$ and a certain matrix M.

B. Another possibility is to assume f to be a realization of a Gaussian isotropic random field with mean Ψ and correlation function $R(x,\tilde{x}) = \lambda \exp(-\rho||x-\tilde{x}||)$. Here the function Ψ describes the most plausible form of the unknown regression function, which, as a realization of the random field, may be very different from Ψ but has smoothness properties because of the form of the correlation function (see Gichman and Skorochod (1974)). This description leads to a prior distribution for f^ξ:

$$f^\xi \sim N(\Psi^\xi, W^{-1}) \tag{9}$$

with the matrix $W^{-1} = ((\lambda \exp(-\rho||x_i-x_j||)))$ assumed to be non-singular.

C. With a fixed linear interpolation operator T we may partition

$$f = Tf^\xi + (f-Tf^\xi) = f_n+f_r. \tag{10}$$

f may be assumed to be a realization of a Gaussian field generated by independent random fields f_n and f_r through equation (10), where $f_n = Tf^\xi$ and

$$f^\xi \sim N(\Psi^\xi, W^{-1}), \; Ef_r = \Psi_r = \Psi - T\Psi^\xi, \; f_r^\xi = 0 \quad \text{a.s.} \tag{11}$$

It is not necessary to specify the smoothness properties of f_r by assumptions on its distribution, because (11) will be sufficient for determining a Bayes estimator.

D. It may be assumed, that the regression function f is sufficiently well approximated by some polynomial of low order. More generally it could be assumed, that for a linear subspace L of F

$$F_* = \{f \in F \mid \min_{l \in L} ||f-l|| < c\} \tag{12}$$

Case D. will not be treated in our paper and we refer to papers of Petersen (1973), Läuter (1975) and Hoffmann (1977), which derive minimax estimators under special conditions (see also Humak (1977)).

The next theorem assumes a normal distribution $y \sim N(f^{\xi}, \Lambda)$:

Theorem 2: Under the prior distribution of C. the estimator

$$\tilde{f} = \Psi + T(I+W\Lambda)^{-1}(y-\Psi^{\xi}) \tag{13}$$

is a Bayes estimator.

(13) extends the known structure of the Bayes estimator of the mean f^{ξ} (see Humak (1977)).

Now we introduce a smoothness condition corresponding to A. (Ψ=0) or (in some sense) to B.:

$$F_* = \{f \in F \mid ||f^{\xi}-\Psi^{\xi}||_W^2 < c\} \tag{14}$$

Unfortunately we are only able to find a linear minimax estimator with respect to the modified risk function

$$R(\theta,\tilde{f}) = \sup_{x \in \mathfrak{X}} E_\theta |\tilde{f}(x)-f_n(x)|^2 \tag{15}$$

and under the relatively weak assumption, that for each x∈ there is a sequence $\{g_x^i\}$ of functions $g_x^i \in F$ with

$$\lim_{i\to\infty} (g_x^i, f) = f(x) \qquad (f \in F) \tag{16}$$

Theorem 3: Under the above assumptions and for a fixed linear interpolation operator T the estimator

$$\hat{f} = T[(I+c^{-1}W\Lambda)^{-1}(y-\Psi^{\xi})+\Psi^{\xi}]$$

is minimax in the class of linear estimators

$$f = T(Ay+a).$$

The parametric counterpart of this theorem (the BLUE $\hat{\gamma}$ is linear minimax w.r.t. the risk $\sup_j E_{\theta}|\tilde{\gamma}_j-\gamma_j|^2$) is an obvious consequence of the minimax result of Kuks and Olman (see Humak (1977)), although this has not been stated yet.

3. Kernel estimators

As kernel estimators of the form

$$f_K(x) = \sum_{i=1}^{n} \frac{K(x-x_i)y_i}{\sum_j K(x-x_j)} = \sum_i W(x,x_i)y_i \tag{17}$$

are increasingly popular in the literature, we want to discuss some risk properties of them. The intuitive background or kernel estimators is that they decrease the variability of estimators by smoothing at the expense of some bias, which is expected to be small because of the smoothness of the regression function. As we have seen in the previous sections and as the following theorem also shows, a more directed smoothing may lead to better estimators.

We assume

$$\Lambda = I, \ (f,1) = (f^{\xi})^T A \, 1^{\xi} \tag{18}$$

We use the risk (3) and will compare (17) with the unsmoothed estimator $\hat{f} = Ty$ and an estimator combining unsmoothed estimation and kernel smoothing:

$$f_*(x) = b(x) + \sum_i W(x-x_i)(y_i-b(x_i)), \tag{19}$$

where $b=\hat{f}_B$ is the projection of $\hat{f}$ on a fixed linear subspace B of F defined by

$$\hat{f}_B = T\, C(C^TAC)^-C^TAf^{\xi}, \tag{20}$$

assuming that B is generated by functions $c_1,\ldots,c_p$ and $C = ((c_j(x_i)))$.

We remark that, with exception of some very special cases, the matrix

$$W = ((W(x_j,x_i)))$$

will be nonsymmetric and therefore, the kernel estimator f_K will be inadmissible. A uniformly better estimator is e.g. $\tilde{f} = T(W^TW)^{1/2}y$, which is not a kernel estimator. Analogously the "hybrid" estimator (19) is also inadmissible. For the next theorem we need the following notation:

$$G = C(C^TAC)^-C^TA, \quad Q = I-G \tag{21}$$

$$a = tr[AWW^T-A(G+WQ)(G+WQ)^T], \quad b=tr[A-AWW^t] \tag{22}$$

$$c = tr[A-A(G+WQ)(G+WQ)^T] \tag{23}$$

$$|f| = ||f^{\xi}-Wf^{\xi}||_A \tag{24}$$

If f_c denotes the constant function with minimal distance $||f-f_c||$, then we have $|f| = |f-f_c|$ and therefore the norm $|f|$ measures the variability of the function f.

<u>Theorem 4:</u> 1. f_K is better than $\hat{f}$, if $|f-f_C|<b$.

2. f_* is better than $\hat{f}$, $|f-f_B|<c$. (25)

3. f_* is better than f_K, if $|f-f_B|<(|f_B|^2+a)/2|f_B|$

We remark, that the eigenvalues of the matrix WW^T are in the interval [0,1] and that at least one of them will be smaller than one, if the kernel estimator (17) is to be considered as smoothing. Consequently, the trace b will be positive and 1. may be interpreted in the sense, that a kernel estimator will be better than the unsmoothed estimator $\hat{f}$, if the unknown regression function has a sufficiently small variability. The unsmoothed estimator $\hat{f}$ and the kernel estimator f_K will be improverd by the hybrid estimator f_*, if f may be well approximated by a function of B (and if this function has a sufficiently large norm $|f_B|$).

REFERENCES

BUNKE, H. (1973), Approximation of regression functions. Math. Operationsforsch. Statist. 4, 314-325.

BUNKE, O. (1983), Optimal estimation of regression functions based on model choice procedures. Proceedings of the Third Prague Symposium on Asymptotic Statistics.

COLLOMB, G. (1981), Estimation non-parametrique de la regression: revue bibliographique. Internat. Statist. Rev. 49, 75-93.

GIDHMAN, I.I. and SKOROCHOD, A.W. (1974), The theory of stochastic processes. Springer, Berlin.

HOFFMANN, K. (1977) Robust alternatives of the Least Squares Estimator. Math. Operationsforsch. Statist., Ser. Statistics 8, 305-311.

LAUTER, H. (1975), Linear minimax estimation for inadequate models for the I. Petersen approach. Math. Operationsforsch. Statist. 6, 769-774.

HUMAK, K.M.S. (1977), Statistische Methoden der Modellbildung, Band 1, Statistische Inferenz für lineare Parameter. Akademie-Verlag, Berlin.

KARSON, M.J., MANSON, A.R. and HADER, R.J. (1969), Minimum bias estimation and experimental design for response surfaces. Technometrics 11, 461-475.

KIMELDORF, G. and WAHBA, G. (1970), A correspondence between Bayesian estimation on stochastic processes and smoothing by splines. Ann. Math. Statist. 41, 495-502.

MONTGOMERY, D.C. and PECK, E.A. (1982), Linear regression analysis. Wiley, New York.

PETERSEN, I. (1973), Linear minimax estimation of inadequate models in L²-metric. Math. Operationsforsch. Statist. 4, 463-471.

SEBER, G.A.F. (1977), Linear regression analysis. Wiley, New York.

THRUM, R. (1979), Linear estimation and prediction in stochastic processes. Math. Operationsforsch. Statist., Ser. Statistics 10, 333-346.

APPENDIX

Proof of Theorem 1.

We introduce the notation $P_N := N(0,\Lambda)$, $S := \sup_{f \in F_*}$,

$$r[\lambda,P,\tilde{f}] := \int r(f,P,\tilde{f})d\lambda(f) \quad \text{(for a p.d. } \lambda \text{ over } F_*)$$

and Z for the Gramian matrix with elements (h_i, h_j), assuming that the operator T is given by (5). We remark, that

$$r(\theta, \hat{f}) = ||f_r||^2 + \text{tr } Z\Lambda.$$

The theorem is proved following the usual reasoning (see Humak (1977)), if there is a sequence of p.d.'s λ_i over F with

$$r_i := \min_{\tilde{f}} r[\lambda_i, P_N, \tilde{f}] \quad \text{and}$$

$$\lim_i r_i = s := Sr(f, P_N, \hat{f}) = \sup_\theta r(\theta, \hat{f}) \ . \tag{26}$$

The p.d. λ_i is constructed as the p.d. of $f = Tu+v$, where u and v are independent random variables with $u \sim N(0, iI_n)$ and

$$P\{v = f_r^i\} = P\{v = - f_r^i\} = 1/2$$

and where $\{f^i\}$ is a sequence with $f^i \in F$ and $\lim_i ||f_r^i|| = S||f_r||$. Under these assumptions, the estimator $\tilde{f}_i$ minimizing $r[\lambda_i, P_N, \tilde{f}]$ (i.e. a Bayes estimator w.r.t. λ_i) is obviously given by

$$\tilde{f}_i = TE(u|y) + E(v|y) = TC_i y,$$

where $C_i := (I + i^{-1}\Lambda)^{-1}$. With $K_i := I - C_i$ and $M_i = ZC_i\Lambda C_i^T$ follows

$$\lim_i i\ K_i\ K_i^T = 0,\ \lim_i M_i = Z\Lambda,$$

$$r(f, P_N, \tilde{f}_i) = ||Tf^\xi + f_r - TC_i f^\xi||^2 + \text{tr} M_i, \tag{27}$$

$$\begin{aligned} &\lim_i \int ||Tf^\xi - TC_i f^\xi + f_r||^2 d\lambda_i(f) = \lim_i E||h^T K_i u + v||^2 = \\ &= \lim_f \text{tr } ZK_i(iI)K_i^T + ||f_r^i||^2 = S||f_r||^2 \end{aligned} \tag{28}$$

and finally from (26), (27) and (28):

$$\lim_i r_i = \lim_i r[\lambda_i, P_N, \tilde{f}_i] = S||f_r||^2 + \mathrm{tr}Z\Lambda = s.$$

<u>Proof of theorem 2.</u>

The Bayes estimator is (13) because the p.d. of y does not depend on f_r and therefore

$$\tilde{f} = E(f|y) = T[\Psi^{\xi} + (I+W\Lambda)^{-1}(y-\Psi^{\xi})] + \Psi_r.$$

<u>Proof of theorem 3.</u>

The proof is indirect, assuming, that there is an estimator $\overline{f} = T(Ay+a)$ with

$$\overline{s} := SR(\theta,\overline{f}) < \hat{s} := SR(\theta,\hat{f}).$$

Then there is an $\varepsilon>0$ and an $x\in\mathfrak{X}$ with

$$SE_\theta|\hat{f}(x)-f_n(x)|^2 > \hat{s}-2\varepsilon > \overline{s}. \tag{29}$$

From our assumptions follows for each estimator of the form $\tilde{f} = T\tilde{f}^{\xi} = T(Ay+a)$

$$\rho_i(\theta,\tilde{f}) := E_\theta|(g_x^i,\tilde{f})-(g_x^i,f_n)|^2 = E_\theta||\tilde{f}^{\xi}-f^{\xi}||^2_{H_i} \tag{30}$$

and (as it will be shown later) also $\lim_i V_i = 0$, where $H_i := (((g_x^i,h_j)(g_x^i,h_\ell)))$ and

$$V_i := |S\rho_i(\theta,\tilde{f})-SE_\theta[\tilde{f}(x)-f_n(x)]^2|.$$

This and (29), (30) secure the existence of an i with

$$S\rho_i(\theta,\overline{f}) < SE_\theta|\overline{f}(x)-f_n(x)|^2+\varepsilon \le \overline{s}+\varepsilon < S\rho_i(\theta,\hat{f}). \tag{31}$$

(31) is a contradiction to the minimaxity of the estimator

$$\hat{f}^{\xi} = \Psi^{\xi} + (I + c^{-1}W\Lambda)^{-1}(y - \Psi\xi)$$

of f^{ξ} w.r.t. the risk $E_{\theta}||\tilde{f}^{\xi} - f^{\xi}||^2_{H_i}$ (see (30)) in the class of linear estimators $\tilde{f}^{\xi}$, as proved by Kuks and Olman (see Humak (1977)).

For the proof of $\lim_i V_i = 0$ we remark, that with the notation

$$w := f - f_n = h^T[(Ay+a) - f^{\xi}], \quad p_i := (g^i_x, w),$$

$$q := w(x)$$

obviously holds

$$V_i \leq S|E_{\theta}P_i^2 - E_{\theta}q^2| \leq 2[SE_{\theta}|p_i - q|^2 + SE_{\theta}q^2]^{1/2}. \tag{29}$$

For the random vector $z := Ay + a - f^{\xi}$ we have bounded moments of second order ($SE_{\theta}z^T z < \infty$) because of the boundedness of the set F_* (see (14)). Therefore our assertion follows from (29) and from $\lim_i d_i = 0$ for $d_i := (g^i_x, h) - h(x)$ and $\lim_i SE_{\theta}|p_i - q|^2 =$

$$= \lim_i SE_{\theta}(d_i^T z)^2 = 0, \; SE_{\theta}q^2 = Sh(x)^T E_{\theta}z\, z^T h(x) < \infty.$$

Proof of theorem 4.

The risk of the estimators are easily calculated as

$$r(\theta, \hat{f}) = E_{\theta}||f^{\xi} - y||^2_A = \mathrm{tr}A, \tag{30}$$

$$r(\theta, f_K) = ||f^{\xi} - Wf^{\xi}||^2_A + \mathrm{tr}AWW^T, \tag{31}$$

$$r(\theta, f_*) = ||(I-W)Wf^{\xi}||^2_A + \mathrm{tr}A(G+WQ)(G+WQ)^T \tag{32}$$

Assertions 1. and 2. follow directly from these formulae (e.g. because of $||(I-W)Qf^{\xi}||_A = |f-f_B|$).

3. follows from the inequality

$r(\theta,f_K)-r(\theta,f_*) = a+u^2+2[(I-W)Qf^{\xi}]^T A[(I-W)Qf^{\xi}] \geq a+u^2-2uv>0,$

valid under (25) with the notation $u := |f_B|$, $v := |f-f_B|$.

Olaf Bunke
Humboldt-Universität
Sektion Mathematik
Unter den Linden 6
DDR-1086 Berlin

Proc. of the 4th Pannonian Symp.
on Math. Stat., Bad Tatzmannsdorf, Austria 1983
W. Grossmann, G. Pflug, I. Vincze, W. Wertz, eds

THE BABOONS COME DOWN FROM THE TREES QUITE NORMALLY

SÁNDOR CSÖRGŐ, LAJOS HORVÁTH
Szeged

Wagner and Altmann (1973) reported an analysis of a set of baboon descent times, some of which were left censored due to varying arrival times at the observation site. Ware and DeMets (1976) based their reanalysis directly on the product-limit estimate for the descent time distribution function. We plot this estimate and fit the curve of a completely specified normal distribution function to it. Using recent theory for the analysis of right or left censored data, we test the goodness of this fit by asymptotically distribution-free Kolmogorov-Smirnov and Cramér-von Mises type statistics and find that the significance of the departure from our normal hypothesis is reasonably small, provided that forty minutes of early morning variability have passed.

1. INTRODUCTION. Wagner and Altmann (1973) reported a set of data of the mean descent time of a group of yellow baboons from their sleeping trees in the Masai-Amboseli Game Reserve, Kenya. Some of the data were censored on the left, since the observers frequently arrived too late to witness the descent. The authors proposed a technique for estimating the probability of descent

in a given time-segment, and their estimator was later commented on by Ware and Demets (1976). The latter authors computed the Kaplan-Meier product-limit estimate of the distribution function of the descent time X^o and estimated the mean and variance of X^o and the probabilities $pr(X^o \in I)$ for certain time-intervals I relying directly upon the product-limit estimate. While, as we shall see it, their mean and variance estimation is indeed successful, it is quite clear at the outset that interval probability estimation cannot, in general, be satisfactorily based on an *unsmoothed* product-limit estimator due to its erratic random fluctuations, especially over short intervals and on the tails where we have fewer data. The aim of the present note is to overcome such kind of difficulties in the case of the baboon descent data by providing enough evidence in Section 3 that the descent time follows a specified normal distribution.

2. MATHEMATICAL BACKGROUND. To formulate the corresponding mathematical model, let $X_1^o,\ldots,X_n^o$ be independent random variables (the descent times) with common continuous distribution function $F^o(t) = pr(X^o \le t)$ to be estimated. These are censored on the left by $Y_1,\ldots,Y_n$ (the arrival times of the observer), a sequence of independent random variables with common distribution function H, independent of the X^o sequence. We can only observe the sequence of pairs $(X_1,\delta_1),\ldots,(X_n,\delta_n)$, where $X_k = \max(X_k^o,Y_k)$ and δ_k is the indicator of the event $\{X_k = X_k^o\}$, $k = 1,\ldots,n$, i.e., $\delta_k = 1$ if X_k^o is uncensored and $\delta_k = 0$ otherwise. Rather than reversing the order of the data at each particular application with left censoring as proposed by Ware and DeMets (1976),

Csörgő and Horváth (1980) proposed reversing the whole theory of right censorship. The corresponding left product-limit estimate $\hat{F}_n^o$ of F^o is defined as

$$\hat{F}_n^o(t) = \begin{cases} \text{undefined} & , t<X_{1:n}, \\ \prod_{\{1 \le k \le n \,:\, X_{k:n} > t\}} \left(\frac{k-1}{k}\right)^{\delta_{k:n}} & , X_{1:n} \le t < X_{n:n}, \\ 1 & , t \ge X_{n:n}, \end{cases}$$

where $X_{1:n} \le \ldots \le X_{n:n}$ are the order statistics corresponding to the sample $X_1,\ldots,X_n$ and $\delta_{k:n}$ is the δ corresponding to $X_{k:n}$. Tied elements in this sample are considered as distinct observations with uncensored observations preceding censored ones. The common distribution function of the observed maxima is $F(t) = \text{pr}(X_k \le t) = F^o(t)H(t)$ for all t. Its empirical counterpart is the ordinary empirical distribution function

$$\hat{F}_n(t) = \frac{1}{n}\#\{1 \le k \le n : X_k \le t\},$$

and we also need the empirical sub-distribution function of the uncensored observations

$$\tilde{F}_n(t) = \frac{1}{n}\#\{1 \le k \le n : X_k \le t,\ \delta_k = 1\}.$$

Define the random function

$$d_n(t) = \int_t^\infty \frac{1}{(\hat{F}_n(s))^2}\, d\tilde{F}_n(s) = \frac{1}{n} \sum_{\{1 \le k \le n \,:\, X_k > t\}} \frac{\delta_k}{(\hat{F}_n(X_k))^2},$$

and finally introduce the following weighted and transformed product-limit process

$$\check{Z}_n(t) = \left(\frac{n}{d_n(T)}\right)^{1/2} \frac{\hat{F}_n^o(t)-F^o(t)}{F^o(t)}, \quad t \geq T,$$

where T is a fixed number such that $F(T) > 0$. Now our Kolmogorov--Smirnov statistic for testing the hypothesis that the distribution function of (the descent time) X^o is indeed F^o (at least on $[T,\infty)$) is

$$D_n(T) = \sup_{T \leq t < \infty} |\check{Z}_n(t)|,$$

and the corresponding Cramér-von Mises statistic is

$$\omega_n^2(T) = -\int_T^\infty (\check{Z}_n(t))^2 d\left(\frac{d_n(t)}{d_n(T)}\right)$$

$$= \frac{n}{(d_n(T))^2} \int_T^\infty \left(\frac{\hat{F}_n^o(t)-F^o(t)}{F^o(t)\hat{F}_n(t)}\right)^2 d\tilde{F}_n(t)$$

$$= \frac{1}{(d_n(T))^2} \sum_{\{1 \leq k \leq n \,:\, X_k > T\}} \delta_k \left(\frac{\hat{F}_n^o(X_k)-F^o(X_k)}{F^o(X_k)\ \hat{F}_n(X_k)}\right)^2 .$$

Introducing the notations

$$K_n(x) = pr(D_n(T) \leq x),$$

$$V_n(x) = pr(\omega_n^2(T) \leq x),$$

for the distribution functions of these statistics, we have, as a result of Theorem 4.1 of Csörgő and Horváth (1981 b) and the general discussion in Csörgő and Horváth (1980), that

$$\lim_{n\to\infty} K_n(x) = K(x), \quad -\infty < x < \infty,$$

$$\lim_{n\to\infty} V_n(x) = V(x), \quad -\infty < x < \infty.$$

Here

$$K(x) = \mathrm{pr}\left(\sup_{0\le t\le 1} |W(t)| \le x\right) = \frac{4}{\pi}\sum_{k=0}^{\infty} \frac{(-1)^k}{2k+1} \exp\left(-\frac{\pi^2(2k+1)^2}{8x^2}\right) =$$

$$= \left(\frac{2}{\pi}\right)^{1/2} \sum_{k=-\infty}^{\infty} \int_{(4k-1)x}^{(4k+1)x} \exp\left(-\frac{t^2}{2}\right)dt,$$

$$V(x) = \mathrm{pr}\left(\int_0^1 W^2(t)dt \le x\right) = 2^{3/2} \sum_{k=0}^{\infty} \binom{-\frac{1}{2}}{k}\left\{1-\Phi\left(\frac{4k+1}{2x^{1/2}}\right)\right\}$$

$$= 1 - \frac{2}{\pi}\sum_{k=0}^{\infty} (-1)^k \int_{\pi(2k+\frac{1}{2})}^{\pi(2k+\frac{3}{2})} \frac{\exp(-\frac{1}{2}xt^2)}{4(-\cos t)^{1/2}}\, dt,$$

where W is the standard Brownian motion (Wiener process), Φ is the standard normal distribution function as in the next section, and where the second and third equalities are well known (see Feller (1966), p. 330, and Csörgő and Horváth (1981 a), respectively). The corresponding distribution functions for the Brownian bridge in place of the motion have been extensively tabulated a long time ago, but we could not find any table published in the literature for $V(x)$ and $K(x)$. A detailed table for $V(x)$ is now provided by Csörgő and Horváth (1981 a), while $K(x)$ is tabulated in Table 1 here. For any x, our program used as many terms from the first series as was required to obtain $K(x)$ with a precise fifth decimal. The values were then rounded off to obtain Table 1 with four decimals. We note that the series has proved to converge very rapidly, we never needed the fifth term of it.

3. A NORMAL HYPOTHESIS FOR THE BABOON DESCENT. On Figure 1 we have the product-limit estimator $\hat{F}^{o}_{152}(t)$ based on the 58 uncensored and the 94 censored observations, reported by Wagner and Altmann in minutes, or by Ware and DeMets in hours. The estimated mean descent time is $\hat{\mu} = 8.12\text{h}$, and the estimated variance of the descent time is $\hat{\sigma}^2 = 0.78\text{h}$, as in Ware and DeMets (1976).

We took 6h50 min = 6.83h as the origin on the time axis. Thus $\hat{\mu} = 8.12\text{h}$ is 77 min, $\hat{\sigma}^2 = 0.78\text{h} = 2808$ min, $\hat{\sigma} = 0.88\text{h} =$ = 53 min. To the plotted $\hat{F}^{o}_{152}(t)$ we fitted the curve of the distribution function of N(77,3025), the normal distribution with mean 77 min and standard deviation $\sqrt{3025} = 55$ min = 0.92h, i.e., we formulated the null hypotheses for the distribution function F^o of the descent time X^o that

$$H_o \; : \; F^o(t) \equiv \Phi\left(\frac{t-77}{55}\right), \; t \geq T,$$

where

$$\Phi(x) = \frac{1}{\sqrt{2\pi}} \int_{-\infty}^{x} \exp\left(-\frac{t^2}{2}\right) dt.$$

Of course, the significance of departure from H_o should depend heavily on T as it is obvious from Figure 1. In Table 2 we present the results of our computations. The values (representing the significance of departure from H_o above T as measured by the K-S statistic) of the third column were computed by entering Table 1 for K(x), while the quantities of the fifth column (giving the significance of departure corresponding to the C-vM statistic)are given by using our table for V(x) in

Csörgő and Horváth (1981 a). Both statistics show reasonably good fit above T = 40. The C-vM statistic gives small significance even with T ≥ 32. This is because $\omega_n^2(T)$ changes only if T crosses an uncensored observation, while $D_n(T)$ can become large if the distance between two neighboring uncensored observations is not too small (cf. the row T = 32 in Table 2).

In Table 3 we compare the Wagner - Altmann and Ware-DeMets values of estimated probabilities of descent in certain time intervals with our probabilities obtained by hypothesis H_0. On the basis of the test we do not have any ground to claim that the first two normal probabilities (0.11 and 0.14) are good estimates in the pre-7.5h tail. However, we feel that the Ware--DeMets estimate for the probability of descent before 7 o'clock is too large because there have been just 2 descents out of the 58 uncensored ones falling there.

4. DISCUSSION. Our approach to the baboon problem may be questionable on the ground that the statistics $D_n(T)$ and $\omega_n^2(T)$, when applied to the baboon data, are in fact the K-S and C-vM functionals of the process $Z_n^*(t)$ defined by putting $\Phi((t-m_n)/\sigma_n)$ in place of $F^o(t)$ in the definition of $Z_n(t)$, where m_n and σ_n are estimators of the mean and the standard deviation based on the same data set. This would mean that we are testing the composite hypothesis that F^o belongs to the normal family instead of the simple hypothesis H_0. Nothing is known about the convergence of the above Z_n^*.

We note that $\hat{F}^o_{152}(77) = \frac{1}{2}$, i.e., our location estimator $\hat{\mu}_{152} = \int_{-\infty}^{\infty} x d\hat{F}^o_{152}(x) = 77$ is also the median estimator. Also, our estimate 55 for the scale parameter σ may be obtained by

Figure 1

THE PRODUCT - LIMIT ESTIMATOR AND THE FITTED NORMAL DISTRIBUTION FUNCTION N (77 min , 3025 min)

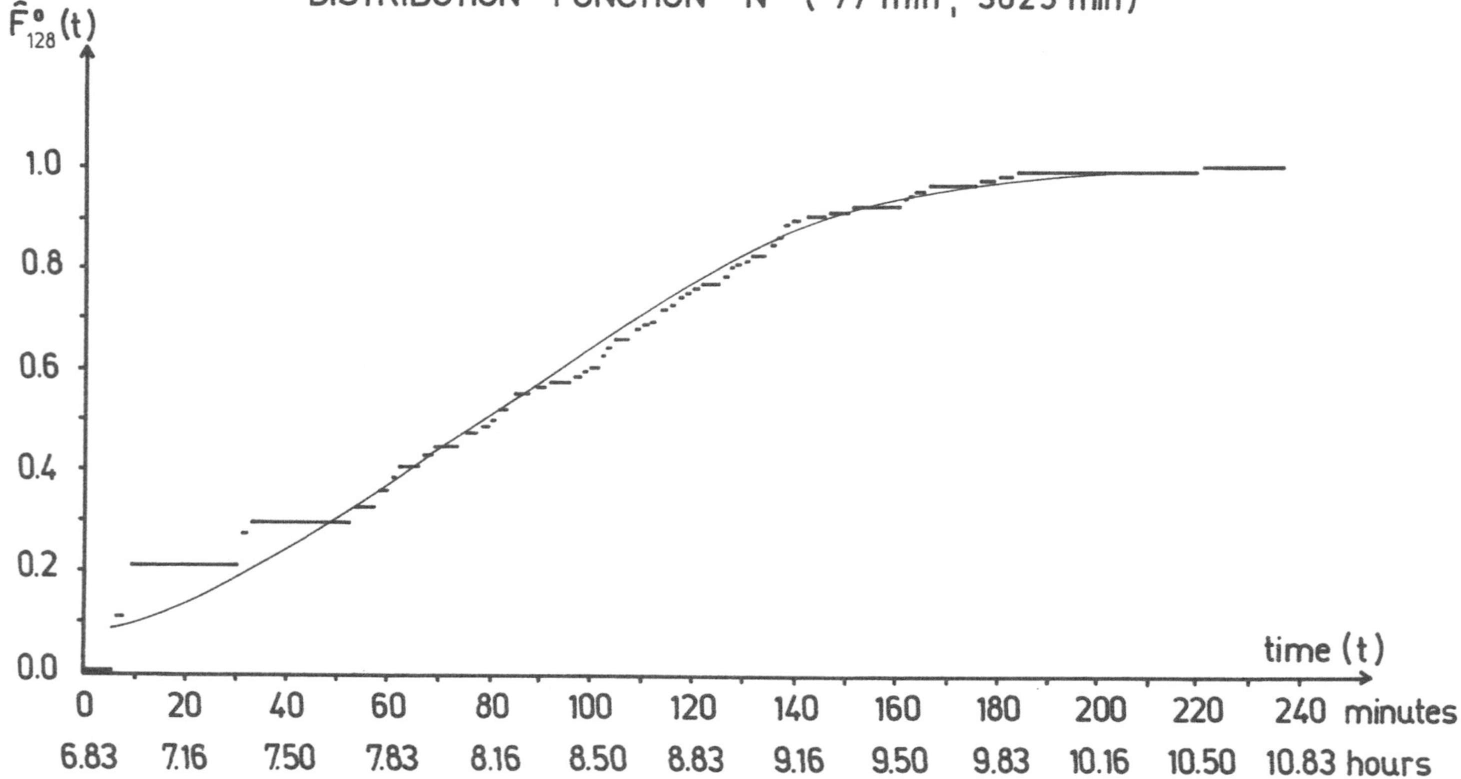

solving the equation

$$\int_{40}^{\infty} x d\Phi \left(\frac{x-77}{\sigma}\right) = \int_{40}^{\infty} x d\hat{F}_{152}^{o}(x) = 74.53,$$

and this differs from the estimated standard deviation by two minutes only. These circumstances were taken as sufficient ground that 77 and 55 are the true values of the parameters. We also investigated the goodness of fit of other normal curves by varying the parameters around 77 and 53. It was the pair (77, 55) at which we obtained the smallest values of the statistics. We thank Miss Kata Kasza for drawing the figure.

Table 1: THE DISTRIBUTION FUNCTION K(x) OF THE ABSOLUTE SUPREMUM OF BROWNIAN MOTION ON [0,1].

	+0.01	+0.02	+0.03	+0.04	+0.05	+0.06	+0.07	+0.08	+0.09	+0.10
0.3	.0000	.0000	.0000	.0000	.0000	.0001	.0002	.0002	.0004	.0006
0.4	.0008	.0012	.0016	.0022	.0029	.0037	.0048	.0060	.0075	.0092
0.5	.0111	.0133	.0158	.0185	.0216	.0249	.0286	.0325	.0368	.0414
0.6	.0462	.0514	.0569	.0626	.0687	.0750	.0815	.0883	.0954	.1027
0.7	.1102	.1179	.1257	.1338	.1420	.1504	.1589	.1676	.1764	.1852
0.8	.1942	.2033	.2124	.2216	.2309	.2401	.2495	.2588	.2682	.2776
0.9	.2870	.2964	.3058	.3152	.3245	.3338	.3431	.3524	.3616	.3708
1.0	.3799	.3890	.3980	.4069	.4158	.4247	.4334	.4421	.4507	.4593
1.1	.4677	.4761	.4844	.4927	.5008	.5089	.5169	.5248	.5326	.5404
1.2	.5480	.5556	.5631	.5704	.5778	.5850	.5921	.5992	.6061	.6130
1.3	.6198	.6265	.6331	.6396	.6461	.6524	.6587	.6649	.6710	.6770
1.4	.6830	.6888	.6946	.7003	.7059	.7114	.7169	.7223	.7276	.7328
1.5	.7380	.7430	.7480	.7529	.7577	.7625	.7672	.7718	.7763	.7808
1.6	.7852	.7895	.7938	.7980	.8021	.8062	.8102	.8141	.8180	.8217
1.7	.8255	.8291	.8327	.8363	.8398	.8432	.8465	.8498	.8531	.8563
1.8	.8594	.8625	.8655	.8685	.8714	.8742	.8770	.8798	.8825	.8851
1.9	.8877	.8903	.8928	.8952	.8976	.9000	.9023	.9046	.9068	.9090
2.0	.9111	.9132	.9153	.9173	.9193	.9212	.9231	.9249	.9268	.9285
2.1	.9303	.9320	.9337	.9353	.9369	.9385	.9400	.9415	.9430	.9444
2.2	.9458	.9472	.9485	.9498	.9511	.9524	.9536	.9548	.9560	.9571
2.3	.9582	.9593	.9604	.9614	.9625	.9635	.9644	.9654	.9663	.9672
2.4	.9681	.9690	.9698	.9706	.9714	.9722	.9730	.9737	.9745	.9752
2.5	.9759	.9765	.9772	.9778	.9785	.9791	.9797	.9802	.9808	.9814
2.6	.9819	.9824	.9829	.9834	.9839	.9844	.9848	.9853	.9857	.9861
2.7	.9865	.9869	.9873	.9877	.9881	.9884	.9888	.9891	.9895	.9898
2.8	.9901	.9904	.9907	.9910	.9913	.9915	.9918	.9920	.9923	.9925
2.9	.9928	.9930	.9932	.9934	.9936	.9938	.9940	.9942	.9944	.9946

3.0	.9948	.9949	.9951	.9953	.9954	.9956	.9957	.9959	.9960	.9961
3.1	.9963	.9964	.9965	.9966	.9967	.9968	.9970	.9971	.9972	.9973
3.2	.9973	.9974	.9975	.9976	.9977	.9978	.9978	.9979	.9980	.9981
3.3	.9981	.9982	.9983	.9983	.9984	.9984	.9985	.9986	.9986	.9987
3.4	.9987	.9987	.9988	.9988	.9989	.9989	.9990	.9990	.9990	.9991
3.5	.9991	.9991	.9992	.9992	.9992	.9993	.9993	.9993	.9993	.9994
3.6	.9994	.9994	.9994	.9994	.9995	.9995	.9995	.9995	.9995	.9995
3.7	.9996	.9996	.9996	.9996	.9996	.9996	.9997	.9997	.9997	.9997
3.8	.9997	.9997	.9997	.9998	.9998	.9998	.9998	.9998	.9998	.9998

Table 2: VALUES OF THE STATISTICS WITH SIGNIFICANCE LEVELS

T (min)	$D_{152}(T)$	$K(D_{152}(T))$	$\omega^2_{152}(T)$	$V(\omega^2_{152}(T))$
70	0.648	0.067	0.076	0.089
60	0.524	0.014	0.080	0.109
50	0.424	0.001	0.027	0.055
40	1.016	0.385	0.027	0.055
32	2.290	0.955	0.027	0.055
30	1.753	0.841	1.035	0.871
20	0.753	0.841	1.035	0.871
6	0.929	0.303	0.277	0.332

Table 3: COMPARISON OF INTERVAL PROBABILITY ESTIMATES

Time period	Wagner-Altmann probabilities	Ware-DeMets probabilities	Normal probabilities
- 7.0	0.34	0.24	0.11
7.0 - 7.5	0.15	0.07	0.14
7.5 - 8.0	0.13	0.15	0.20
8.0 - 8.5	0.17	0.16	0.21
8.5 - 9.0	0.20	0.22	0.17
9.0 - 9.5	0.10	0.10	0.10
9.5 - 10.0	0.06	0.05	0.05
10.0 -	0.01	0.01	0.02

REFERENCES

1 CSÖRGŐ, S. and HORVÁTH, L. (1980): Random censorship from the left. Studia Sci. Math. Hungar. 15, pp. 397-401.

2 CSÖRGŐ, S. and HORVÁTH, L. (1981 a): On the Koziol-Green model for random censorship. Biometrika 68, pp. 391-401.

3 CSÖRGŐ, S. and HORVÁTH, L. (1981 b): On random censorship from the right. Acta Sci. Math. (Szeged) 44, pp. 23-34.

4 FELLER, W. (1966). An Introduction to Propability Theory and its Applications. Vol. 2. Wiley, New York.

5 WAGNER, S.S. and ALTMANN, S.A. (1973). What time do the baboons come down from the trees? (An estimation problem). Biometrics 29, pp. 623-635.

6 WARE, J.H. and DeMETS, D.S. (1976). Reanalysis of some baboon descent data. Biometrics 32, pp. 459-463.

Sándor Csörgő, Lajos Horváth
Bolyai Institute
Szeged University
Aradi Vértanúk tere 1.
H-6720 Szeged

Proc. of the 4th Pannonian Symp. on Math. Stat., Bad Tatzmannsdorf, Austria 1983
W. Grossmann, G. Pflug, I. Vincze, W. Wertz, eds

ROBUST AND BOUNDED INFLUENCE REGRESSION: REALIZATION AND APPLICATION

R. DUTTER

Graz

ABSTRACT

Bounded Influence Regression has received much attention in recent years. It should provide results which are stable in respect to the additive errors (distributional robustness) as well as in respect to outliers in the factor space (leverage points, influential observations).

We discuss some realizations of the solution of this problem which includes estimation in the general linear model (generalized M-estimates and testing of the linear hypothesis (robust F-tests). The practical implementation in a widely used regression computer program (BLINWDR) and an illustrative example are presented.

1. Introduction

Ordinary least-squares analysis of the classical linear regression model is efficient only if the appropriate assumptions are valid. Even small departures from the standard assumptions on the model may seriously damage the results, that means, they may cause inefficiency, bias, inconsistency, and incorrect tests of significance.

Robust and bounded influence regression techniques have been introduced to overcome these difficulties. The basic idea is the weighting of influential observations. We only mention the paper by Hampel (1978) who used the infinitesimal approach for finding robust estimation techniques which bound the influence of outlying observations, and Krasker and Welsch (1982) who also provide some historical notes. Recently, Huber (1983) gives a critical discussion from the point of view of the minimax approach.

Tests of the linear hypothesis were robustified e.g. by Schrader and Hettmansperger (1980), in order to manage more robust loss functions than least-squares. A considerable advance for the bounded influence regression problem was achieved by E. Ronchetti in his thesis whose results were partially presented in Ronchetti (1982).

For computing bounded influence regression, usually a high-speed computer is necessary, and because of the many possibilities, considerably much output is produced. Marazzi (1980) presented a collection of subprograms for the estimation problem. Unfortunately, the user has to write his own main program, and links between the user's data and all the subroutines are not always easy establish. It is mainly meant for statisticians highly experienced in programming. Only very recently (see Marazzi et al., 1983) he reports a project for creating an "easy to use" package intended for a broader user population.
At the COMPSTAT-meeting Samarov and Welsch (1982) presented a computer program which runs on the TROLL-system and which seems to be hard to implement on other systems. At the same meeting (see Dutter, 1982) a preliminary version of a user-oriented and portable program was presented. This computer program (named BLINWDR) now contains many options for computing bounded influence estimates and certain test statistics for linear hypotheses. It is an extension of a widely used linear regression program with all its options on variable choice, transformations, plots, etc.

The aim of the present paper is not the exhaustive treatment of bounded influence regression but rather the presentation and illustration of the notions and options realized in a computer program. We mention the general features in the next section, introduce some possible estimation and test procedures in Sections 3 and 4, and give an illustration with real data in Section 5. The FORTRAN program is available on request from the author.

2. The Computer Program BLINWDR

The robust and bounded influence regression analysis is embedded in the well-known "Linear Least-squares Curve Fitting Program LINWOOD" which is described in Daniel and Wood (1980). It essentially fits the model

$$f(\underset{\sim}{\theta}) = \theta_o + \theta_1 x_1 + \ldots + \theta_k x_k$$

with $\theta_o, \ldots, \theta_k$ k+1 unknown parameters to n observations $y_1, \ldots, y_n$ by the classical method of least-squares

$$\sum_{i=1}^{n} (y_i - f_i(\underset{\sim}{\theta}))^2 w_i = \min. \qquad (2.1)$$

where $f_i(\underset{\sim}{\theta}) = \theta_o + \theta_1 x_{i1} + \ldots + \theta_k x_{ik} = \underset{\sim}{\theta}^T \underset{\sim}{x}_i$ contains the i-th values of the independent variables $x_1, \ldots, x_k$. w_i denotes a user-defined (known) weight (usually = 1). The program offers the standard statistical information and many additional options in respect to computation and output (see e.g. Dutter, 1982).

The program BLINWDR contains extensions of the least-squares fitting to robust and bounded influence regression fitting and, correspondingly, of the statistics and tables with interesting information.

We discuss briefly some methods in the following two sections. A full description of all possibilities of the program is given in a technical report (Dutter, 1983b).

3. Estimation in Robust and Bounded Influence Regression Analysis

In regression analysis we are concerned with two types of outliers, one is characterized by a large residual (outlier in the y-variable, "in distribution"), the second one by a "strange" value in the factor (x-) space, which may influence largely the analysis.

In order to limit the influence of these outliers the following estimation procedure instead of least-squares (2.1) has been proposed. Solve

$$\sum_{i=1}^{n} \rho\left(\sqrt{w_i}\,\frac{y_i - f_i(\underset{\sim}{\theta})}{\sigma p_i}\right) v_i\, p_i = \min.$$

for $\underset{\sim}{\theta}$. The function ρ should limit the influence of large residuals (outliers in distribution) and should therefore be a function which goes less fast to infinite than the square function. The weights p_i and v_i are introduced in order to bound the influence of outliers in the factor space, and only depend on the values in factor space ($\underset{\sim}{x}_i$, say). If all weights are 1, i.e. $p_i \equiv v_i \equiv 1$ for all i, we obtain a <u>Huber-type</u> estimator, if $p_i \equiv 1$ a <u>Mallows-type</u> estimator and if $p_i \equiv v_i$ a <u>Schweppe-type</u> estimator. σ denotes a scaling factor which also has to be calculated from the data. In case of convex ρ we propose to solve it by a sort of Huber-proposal-II, which may be written as general minimum problem in $\underset{\sim}{\theta}$ and σ by

$$\sum_{i=1}^{n} \rho\left(\sqrt{w_i}\,\frac{y_i - f_i(\underset{\sim}{\theta})}{\sigma p_i}\right) v_i p_i \sigma + A\sigma = \min.$$

with an appropriate constant A such that the estimate of σ^2 becomes asymptotically unbiased for normal errors.

The program BLINWDR offers the choice of several ρ-functions. We note three of them by their derivatives Ψ :

(a) $\Psi_{LS} = t$ (least-squares),

(b) $\Psi_H(t) = \max\{-c, \min(t,c)\}$ with $c > 0$ (Huber's Ψ),

(c)
$$\Psi_{HA}(t) = \begin{cases} t & |t| \le a \\ a.\mathrm{sgn}(t) & a < |t| \le b \\ \dfrac{d-|t|}{d-b}\, a.\mathrm{sgn}(t) & b < |t| \le d \\ 0 & d < |t| \end{cases}$$

with $0 < a < b < d$ (Hampel's Ψ).

The weights p_i and v_i are functions of the size of $\underset{\sim}{x}_i$ in a metric defined by a (robust) covariance structure of the factor space. We report briefly the idea (see Hampel, 1978, and Dutter, 1983a, for details). Find a transformation matrix A such that

$$\frac{1}{n}\sum_{i=1}^{n} u(||\underset{\sim}{z}_i||)\underset{\sim}{z}_i\underset{\sim}{z}_i^{T} = \gamma I.$$

where $\underset{\sim}{z}_i = A\underset{\sim}{x}_i$. (u and γ are specified below.) The size of $\underset{\sim}{x}_i$ is then defined by the euclidean norm of $\underset{\sim}{z}_i$, i.e. $||\underset{\sim}{z}_i||$, and the weights p_i and v_i are certain functions of $||\underset{\sim}{z}_i||$. The program has the following options:

(a) Mallows-type estimator, Hampel's proposal (bounded influence on the estimated observations):
$u(t) = \min\{1, \mathrm{BOUND}/t^2\}$, $v_i = u(||\underset{\sim}{z}_i||)$, $\gamma = 1$.

(b) Mallows-type estimator, Mallows/Maronna's proposal (bounded influence on the estimated parameters): $u(t) = \min\{1, \text{BOUND}/t^2\}$, $v_i = \sqrt{u(||\underset{\sim}{z}_i||)}$, $\gamma = \frac{1}{k+1}\xi[u(||\underset{\sim}{z}||)||\underset{\sim}{z}||^2]$.

(c) Schweppe-type estimator, Hampel/Krasker's proposal (bounded total influence): $u(t) = 2\Phi(\delta/t)-1$, $v_i = \min\{1, 1/||z_i||\}$, $\gamma = 1$, where δ is a certain constant and Φ stands for the standard normal distribution function.

(d) Schweppe-type estimator with his original proposal: $v_i = \sqrt{1-h_{ii}}$ where h_{ii} denotes the i-th diagonal element of the hat matrix $X(X^TWX)^{-1}X^TW$ with the design matrix X and the diagonal matrix W containing the weights w_i.

4. Testing the Linear Hypothesis

Denote the usual ("full") linear model

$$y_i = f_i(\underset{\sim}{\theta}) + \varepsilon_i, \; i = 1,\dots,n,$$

by (Ω) and suppose that the linear hypothesis

$$\ell_j(\underset{\sim}{\theta}) = 0, \; j = q+1,\dots,k,$$

is to be tested, where ℓ_j denote k-q linearly independent functions. Then, through a transformation of the parameter space, the hypothesis may be rewritten as

$$H_o : \theta_{q+1} = \dots = \theta_k = 0.$$

Denote by (ω) the "reduced" model which is imposed by this hypothesis.

It is classically tested with the statistic

$$F = \frac{1}{k-q} \sum_{i=1}^{n} \left\{ \left(\frac{r_i(\omega)}{\hat{\sigma}} \right)^2 - \left(\frac{r_i(\Omega)}{\hat{\sigma}} \right)^2 \right\}$$

where

$$\hat{\sigma}^2 = \frac{1}{n-k-1} \sum (r_i(\Omega))^2$$

and $r_i(\Omega)$ and $r_i(\omega)$ denote the residuals arising from the full and the reduced model, respectively. Under the usual assumption of normally distributed errors, this statistic is distributed $F_{k-q,n-k-1}$.

Ronchetti (1982) has introduced a new class of tests, called the τ-test, which can be viewed as an extension of the log-likelihood ratio test for linear models. It is defined by a test statistic which is a natural generalization of the F-statistic defined above, namely by

$$S_n^2 = S_n^2(\underset{\sim}{x}_1, \underset{\sim}{x}_2, \ldots, \underset{\sim}{x}_n; y_1, \ldots, y_n)$$

$$= \frac{2}{k-q} \frac{1}{n} \sum_{i=1}^{n} \left\{ \tau\left(\underset{\sim}{x}_i(\omega), \sqrt{w_i} \frac{r_i(\omega)}{\sigma}\right) - \tau\left(\underset{\sim}{x}_i(\Omega), \sqrt{w_i} \frac{r_i(\Omega)}{\sigma}\right) \right\}$$

where

$$\tau\left(x_i(.), \sqrt{w_i} \frac{r_i(.)}{\sigma}\right) = \rho\left(\sqrt{w_i} \frac{r_i(.)}{\sigma p_i(.)}\right) v_i(.) \, p_i(.)$$

and the point in brackets (.) emphasizes the dependence on the appropriate model (Ω) or (ω). Ronchetti shows that S_n^2, under H_o, is asymptotically distributed as a linear combination of $k-q$ χ_1^2 -variables.

In a very general context it will probably be difficult to find quantiles of this distribution. We can, however, extract some special cases which are easier to handle and we mention two of them. If q = k-1 we only have one χ^2_1 -variable. But this is a frequent case in variable search procedures, e.g. in stepwise regression.

Another special case appears when the weights from the factor space are all equal to one, i.e. $\tau(\underset{\sim}{x},.) = \rho(.)$. This case of testing the linear hypothesis which corresponds to the robust estimation problem of Huber (1973), is also treated by Schrader and Hettmansperger (1980) (see also Sen, 1982). They have shown that, under certain conditions,

$$F_M = \frac{2}{\lambda(k-q)} \Sigma \{\rho(\sqrt{w_i} \frac{r_i(\omega)}{\sigma}) - \rho(\sqrt{w_i} \frac{r_i(\Omega)}{\sigma})\},$$

where λ is a correction factor ($\lambda = \xi\Psi^2 / \xi\Psi'$), is asymptotically distributed as $F_{k-q,n-k-1}$. In the next section we will illustrate this procedure, and therefore, we try to press it and the output of our program in a usual analysis-of-variance (ANOVA) table scheme.

Table 1 displays an excerpt of an output (the data is explained in the next section) of the computer program BLINWDR. Some important values are underlined. For the ANOVA-table shown in Table 2 we need the "Sum of Tau of Residuals" which is defined by

$$STR = 2 \sum_{i=1}^{n} \tau(\underset{\sim}{x}_i, \sqrt{w_i} \frac{r_i}{\sigma}),$$

the "Residual Sum of Squares (Robustified)",

$$RSSR = \frac{1}{\beta} \sum_{i=1}^{n} \left[\Psi\left(\sqrt{w_i}\ \frac{r_i}{\sigma}\right) \sigma \right]^2$$

with $\beta = \Psi^{2}$ (a correction factor), the "Residual Mean Square (Robustified)", RMSR = RSSR/(n-k-1),

Tab.1: Sample Output of Program BLINWDR

```
PROGRAM  B L I N W D R FOR ROBUST AND BOUNDED INFLUENCE REGRESSION

           DUST DATA: DEP VAR 1 DUST  MIN Y = .390 MAX Y =4.800

ROBUST FIT NO.: 1 (PSI BENDS AT C = 1.345).
HUBER TYPE ESTIMATOR.

IND.VAR(I)  NAME   COEF.B(I)  S.E.COEF.  T-VALUE  R(I)SORD
   0               -20.346      2.6        7.8
   1        DENS     13742.   1.66+003     8.3       .0487
   2          TL-C -.45625     .15         3.1       .4617
   3          TL-L  .29618    5.39-002     5.5       .4811

NO. OF ITERATIONS FOR ROBUST FIT 7    (BETA= .7102)
COMPUTED SCALE SIG = .56679

NO.OF OBSERVATIONS                          33
NO.OF IND. VARIABLES                         3
RESIDUAL DEGREES OF FREEDOM                 29
F-VALUE                                      25.04
F-VALUE (ROBUSTIFIED)                        37.25
F-VALUE (TAU)                                21.44  (LAMDA=.86460)
RESIDUAL ROOT MEAN SQUARE                 .60954066
RESIDUAL ROOT MEAN SQUARE (ROBUSTIFIED)   .56669585
ROOT MEAN TAU OF RESIDUALS                .56851291
RESIDUAL MEAN SQUARE                      .37153982
RESIDUAL MEAN SQUARE (ROBUSTIFIED)        .32114418
MEAN TAU OF RESIDUALS                     .32320693
RESIDUAL SUM OF SQUARES                   10.774655
RESIDUAL SUM OF SQUARES (ROBUSTIFIED)     9.3131813
SUM OF TAU OF RESIDUALS                   9.3730011
TOTAL SUM OF SQUARES                      38.683132
TOTAL SUM OF SQUARES (ROBUSTIFIED)        45.197859
TOTAL SUM OF TAU                          38.683132
MULT.CORREL.COEF.SQUARED                     .7242
MULT.CORREL.COEF.SQUARED (ROBUSTIFIED)       .7940
MULT.CORREL.COEF.SQUARED (TAU)               .7662
```

and the "Residual Degrees of Freedom" n-k-1. Then the test statistic is defined, after two runs of the program, once with the full model (ω), by

$$F_M = \frac{STR(\omega)-STR(\Omega)}{RMSR(\Omega)\ (k-q)}\frac{1}{\lambda}.$$

Tab. 2: ANOVA-table

Source of Variation	STR	DF	MTR	
Full Model (Ω)	STR (Ω)	n-k-1		
Hypothesis H_o	$STR(\omega)-STR(\Omega)$	k-q	$\frac{STR(\omega)-STR(\Omega)}{k-q}$	F
Reduced Model (ω)	STR (ω)	n-q-1		
"Error"	RSSR (Ω)	n-k-1	RMSR(Ω)	

5. An Application

In industrial cardboard production, data has been collected in order to find variables which are responsible for dust generation during the cutting of cardboards. Besides the quantity of produced dust many characteristics of the cardboards, like "weight", "moistness", "glue", "ash", "thickness","density", "tear length-lengthwise", "tear length-crosswise", "stiffness-lengthwise", "stiffness-crosswise", and of the cutting knife, like "mean age", "mean pressure", depth of entry", were recorded. After a detailed study the conclusion was that the data indicate three variables which are most characteristic for the amount of generated dust. They are density, tear length-crosswise (TL-C) and tear length-lengthwise (TL-L). For illustration we now only consider these three variables and display their values together with those of the amount of dust in Table 3.

As an illustration Table 1 shows some results of the Huber-type estimator. If we want to know whether the variable TL-C is of significant influence, i.e. we would test the hypothesis

$H_o : \theta_2 = \text{'TL-C'} = 0,$

then the regression computations have to be redone, normally in the same computer run by reusing the same data set and deleting one variable in the second problem, and with the same scale σ. In Table 4 we display the results of the tests in form of an ANOVA-table where three different Ψ-functions were used. They are indicated by LS for least-squares, by H for Huber and by HA for Hampel. The parameters of the functions Ψ were chosen to be $c = 1.345$ for Ψ_H and $a = 1.345$, $b = 2.0$, $d = 2.7$ for Ψ_{HA}.

Tab. 3: Dust Data

No.	Ident.	Dens.	TL-C	TL-L	Dust
1	90	.00140	8.45	21.85	.92
2	91	.00138	8.50	22.35	1.53
3	92	.00142	8.40	22.00	2.50
4	93	.00143	8.10	21.75	1.89
5	94	.00139	8.35	21.80	1.40
6	95	.00144	8.15	21.85	1.70
7	96	.00132	8.20	25.45	1.80
8	97	.00135	8.40	25.95	2.10
9	98	.00136	8.35	25.70	2.00
10	99	.00134	8.35	26.50	2.00
11	14	.00151	6.35	19.00	2.44
12	15	.00141	6.60	19.10	2.11
13	16	.00150	6.85	18.65	2.89
14	20	.00145	6.90	19.10	1.83
15	21	.00151	6.80	19.10	2.00
16	22	.00149	7.70	20.20	2.25
17	10	.00134	5.75	19.70	1.11
18	11	.00135	6.15	18.45	.89
19	12	.00136	5.95	17.70	.72
20	13	.00136	6.10	17.75	.94
21	8	.00134	5.80	18.30	.83
22	9	.00133	6.00	19.05	.61
23	30	.00144	7.35	19.65	3.10
24	31	.00148	7.55	19.25	2.90
25	32	.00148	6.90	19.25	2.60
26	34	.00146	6.40	20.80	4.80
27	35	.00153	6.55	20.60	4.50
28	80	.00140	7.40	19.55	.44
29	81	.00138	7.10	18.45	.39
30	82	.00138	7.95	18.40	.72
31	83	.00135	7.80	17.30	.50
32	84	.00139	7.60	18.45	.56
33	85	.00140	7.20	18.80	.56

All three "F-values" of the tests turn out, of course, to be significantly large. However, they differ quite remarkably. If we remove one observation, say point no.26, which has a residual about 3 times larger than the estimated standard error, and perform the same tests, we get the results of Table 5. The "F-values" are much closer to each other.

Tab. 4: Test with Data of Table 3

$H_o : \theta_2 = \text{'TL-C'} = 0.$

Source		STR	DF	MTR	
Ω	LS	10.63066			1
	H	9.37300	29		λ = .8646
	HA	8.45810			.8839
H_o	LS	3.92554			10.710
	H	2.39356	1		F = 8.620
	HA	1.41880			5.544
ω	LS	14.55620			
	H	11.76656	30		
	HA	9.87690			
"Error"	LS	10.63066		.366574	
	H	9.31318	29	.321144	
	HA	8.39650		.289534	

Tab.5: Test with Data of Table 3, One Outlier (Point No.26) Removed

$H_o : \theta_2 = \text{'TL-C'} = 0.$

Source		STR	DF	MTR	
Ω	LS	7.27407			1
	H	6.93088	28		λ = .8646
	HA	6.85837			.8339
H_o	LS	1.89640			7.300
	H	1.54777	1		F = 6.478
	HA	1.21337			5.333
ω	LS	9.17047			
	H	8.47865	29		
	HA	8.07174			
"Error"	LS	7.27407		.259788	
	H	7.73807	28	.276360	
	HA	7.63994		.272855	

For ease of illustration we now use a certain subset of our data set and display the standardized values (arithmetic mean subtracted and divided by the emirical standard deviation) in Table 6. First examination of this table (in a univariate manner) only reveals one strange measurement in the amount of dust. At this first view the table does not show any influential point in the factor space. However,

Tab.6: Some Standardized Values of Density, Tear Length-crosswise, Tear Length-lengthwise and Amount of Dust

No.	Dens.	TL-C.	TL-L	Dust
1.	-1.700505	0.796900	1.615319	-0.451600
2.	-1.369433	-1.883120	-1.100765	-0.746390
3.	-1.369433	0.964401	2.014185	-0.292255
4.	-1.203897	1.020235	1.805255	-0.212582
5.	-1.038362	-1.715619	-1.328689	-0.834030
6.	-1.038362	0.964401	1.710287	-0.292255
7.	-0.707290	1.131903	0.437716	-0.666717
8.	-0.541754	0.964401	0.228787	-0.770292
9.	-0.376218	1.076069	0.247780	-1.152722
10.	-0.045146	1.020235	0.304761	0.106110
11.	0.120390	0.685233	0.209793	-0.379895
12.	0.285926	0.741066	0.247780	-0.531273
13.	0.285926	-0.152274	-0.587938	0.584147
14.	0.451462	-0.654778	-0.398002	3.428470
15.	0.451462	-0.654778	-0.796868	-0.427699
16.	0.616997	-1.213115	-0.151085	1.938587
17.	0.948069	0.071061	-0.739887	0.424801
18.	0.948069	-0.654778	-0.739887	0.185783
19.	1.113605	0.238562	-0.379009	-0.093072
20.	1.279141	-0.710612	-0.967810	0.416834
21.	1.444677	-1.268949	-0.834855	0.058306
22.	1.444677	-0.766445	-0.796868	-0.292255

if we draw a two-dimensional picture of Tear Length-crosswise over Density, as shown in Figure 1 (scales of the axes are not standardized), we see two points, named 2 and 5, clearly separated from the bulk of the data. We can compute "distances" of the points from the center of the cloud by calculating first a covariance structure, which is shown by a tolerance ellipse.

It is obvious that a robustly calculated covariance would pronounce much more these outliers (which is very important for the weights of influential points). This is visualized by the dotted line. We have also marked the point number 14 which is an outlier in the y-variable only. A plot of TL-L over the density (not shown here) is similar.

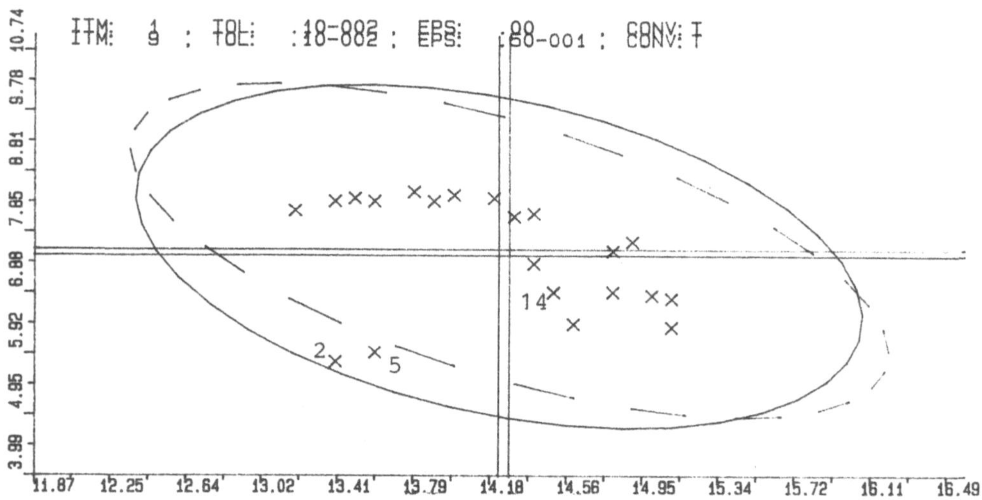

Fig. 1: Tear Length-crosswise over Density ($*10^4$)

To this reduced data set we have applied different estimators, i.e. the Huber-type, the Mallows- and the Schweppe-type estimator with different proposals for weights, and different loss functions (least-squares LS and Huber's Ψ). Some selected results of the weights (p_i or v_i), $\underset{\sim}{\theta}$, σ, estimated observations

$\hat{y}_i$ and residuals r_i are given in Table 7. We focuse on the outliers in factor space y_2 and y_5 and the outlier in distribution y_{14}. It is obvious that the residual of point 14 is more pronounced if Huber's Ψ is used. This is remarked in all types of estimators.

In order to illustrate the behavior of the robust and bounded influence estimators, we have added an artificial outlier in y at an outlier in factor space. More precisely, we have changed the observation y_2 (observation of amount of generated dust) to $y_2 = -3.5$. The selected values of Table 7 for the modified data set are displayed in Table 8.

Tab.7: Estimation Results from Fitting the Data of Table 5

Type:	Huber-type		Mallows-type				Schweppe-type	
Proposal:			Hampel		Mallows/Maronna		Hampel/Krasker	
Loss f.:	LS	H	LS	H	LS	H	LS	H
p_2 or v_2				.066		.390		.253
p_5 or v_5				.077		.420		.272
p_{14} or v_{14}				1.000		1.000		.723
$\hat{\theta}_1$	0.000	-.149	.215	-.030	.039	-.114	.000	-.136
$\hat{\theta}_2$	.611	.497	.111	.218	.489	.443	.611	.494
$\hat{\theta}_3$	-.645	-.416	-.887	-.533	-.692	-.462	-.645	-.403
$\hat{\theta}_4$	.724	.527	.488	.339	.635	.499	724	.505
$\hat{y}_2$	-.419	-.627	1.196	.303	-.027	-.400	-.419	-.611
$\hat{y}_5$	-.489	-.652	.973	.208	-.126	-.445	-.489	-.630
$\hat{y}_{14}$	.410	.137	.651	-.283	.460	.190	.410	.150
r_2	-.327	-.119	-1.94	-1.05	-.719	-.346	-.327	-.135
r_5	-.345	-.182	-1.81	-1.04	-.709	-.389	-.345	-.205
r_{14}	3.018	3.391	2.777	3.146	2.968	3.238	3.018	3.278
$\hat{\sigma}$	.899	.535	.830	.632	.890	.592	.899	1.049

Tab. 8: Estimation Results from Fitting the Data of Table 5 with an Additional Outlier.

Type:	Huber-type		Mallows-type				Schweppe-type	
Proposal:			Hampel		Mallows/Marónna		Hampel/Krasker	
Loss f.:	LS	H	LS	H	LS	H	LS	H
p_2 or v_2			.066		.390		.253	
p_5 or v_5			.077		.420		.272	
p_{14} or v_{14}			1.000		1.000		.723	
$\hat{\theta}_1$	-.125	-.201	.159	-.030	-.067	-.148	-.125	-.159
$\hat{\theta}_2$	1.014	-.724	.284	.218	.827	.542	1.014	.555
$\hat{\theta}_3$	-.438	-.330	-.811	-.533	-.541	-.415	-.438	-.362
$\hat{\theta}_4$	.960	.676	.604	.339	.858	.564	.960	.536
$\hat{y}_2$	-1.74	-1.31	.633	.303	-1.13	-.729	-1.71	-.828
$\hat{y}_5$	-1.70	-1.28	.453	.208	-1.14	-.747	-1.70	-.827
$\hat{y}_{14}$	.238	.073	.578	.283	.319	.144	.238	.115
r_2	-1.76	-2.19	-4.13	-3.80	-2.38	-2.77	-1.76	-2.67
r_5	-.867	.449	-1.29	-1.04	.303	-.087	.867	-.007
r_{14}	3.191	3.355	2.850	3.146	3.110	3.284	3.191	3.314
$\hat{\sigma}$	1.06	.678	.866	.632	.988	.611	1.061	1.076

The most remarkable result is delivered by the Mallows-type estimator, Hampel's proposal, where the values of the estimated parameter $\underset{\sim}{\theta}$ remain the same, with or without the outlier in the dependent variable at an outlying point in the factor space.

REFERENCES

DANIEL C. and F. WOOD (1980): Fitting Equations to Data. J. Wiley, New York.

DUTTER R. (1982): BLINWDR: Robust and Bounded Influence Regression. In: COMPSTAT 1982, Proc. in Comput. Statist., Causinus et al. (eds.), Physica-Verlag, Vienna.

DUTTER R. (1983a): COVINTER: A Computer Program for Computing Robust Covariances and for Plotting Tolerance Ellipses. Res. Rep. No. 10. Institut für Statistik, Technische Universität Graz.

DUTTER R. (1983b): Computer Program BLINWDR for Robust and Bounded Influence Regression. Res. Rep. No. 8. (Version 2). Institut für Statistik, Technische Universität Graz.

HAMPEL F.R. (1980): Optimally Bounding the Gross-Error-Sensitivity and the Influence of Position in Factor Space. Proc. Statist. Comput. Section, Amer. Statist. Assoc., 59-64.

HUBER P.J. (1973): Robust Regression: Asymptotics, Conjectures and Monte Carlo. Ann. Statist. 1, 799-821.

HUBER P. J. (1983): Minimax Aspects of Bounded-Influence Regression. J. Amer. Statist. Assoc., 78, 381.

KRASKER W. S. and R. E. WELSCH (1982): Efficient Bounded-Influence Regression Estimation. J. Amer. Statist. Assoc., 77, 379, 595-604.

MARAZZI A. (1980): ROBETH: A Subroutine Library for Robust Statistical Procedures. In: COMPSTAT 1980, Proc. in Comput. Statist., Barritt et al. (eds.), Physica-Verlag, Vienna.

MARAZZI A., G. BRAOUDAKIS and A. RANDRIAMIHARISOA (1983): Robust Statistical Analysis of Medical Data: A New,Portable, Easy to Use Computer Package. Unpublished paper presented at: Intern. Conf. on Medical Statistics, Sept. 6-9, 1983, Basel, Switzerland.

RONCHETTI E. (1982): Robust Alternatives to the F-test for the Linear Model. In: Prob. and Statist. Inference. 2nd Pann. Symp. Grossmann et al. (eds.), Reidel Publ. Comp., Dordrecht-Holland.

SAMAROV A. and R. E. WELSCH (1982): Computational Procedures for Bounded-Influence Regression. In: COMPSTAT 1982, Proc. in Comput. Statist., Causinus et al. (eds.), Physica-Verlag, Vienna.

SCHRADER R.M. and T.P. HETTMANSPERGER (1980): Robust Analysis of Variance Based Upon a Likelihood Ratio Criterion. Biometrika 67, 93-101.

SEN P.K. (1982): On M-tests in Linear Models. Biometrika 69, 1, 245-8.

ACKNOWLEDGEMENTS

The author likes to thank N. Stockinger for improving and U. Fleischhacker for typing the manuscript. He is also grateful to the referee for his valuable remarks. The research was supported by the "Fonds zur Förderung der wissenschaftlichen Forschung", project no. 4972.

Rudolf Dutter
Institut für Statistik
TU Graz
Hamerlinggasse 6
A-8010 Graz

Proc. of the 4th Pannonian Symp. on Math. Stat., Bad Tatzmannsdorf, Austria 1983
W. Grossmann, G. Pflug, I. Vincze, W. Wertz, eds

MONTE CARLO METHOD FOR NONLINEAR FILTERING

DENIS M. ENĂCHESCU

Bucharest

ABSTRACT

The paper discusses the computational implications of a recently developed reformulation of the optimal-filtering equations for a noisily observed diffusion process. A Monte Carlo method of solution is proposed and properties of the new method are investigated. The efficiency of the algorithm is discussed and a comparision with Kushner's method is given.

1. INTRODUCTION

This paper concerns the filtering problem of calculating the conditional distributions of a diffusion process x_t satisfying a stochastic differential equation of the form

$$dx_t = a(t,x_t)dt + b(t,x_t)dv_t \tag{1}$$

given noisy observation y_t specified by

$$dy_t = c(x_t)dt + dw_t \tag{2}$$

The state vector x_t in equation (1) is m-dimensional, the coefficients a(t,x) and b(t,x) take their values in $\mathbb{R}^m$ and the space of (m,n)-matrices, and v_t is a n-dimensional standard Wiener process with independent components. The noise w_t in equation (2) is a one dimensional Wiener process independent of v_t, and $c(x)$ denotes a real continuous function on $\mathbb{R}^m$. We suppose all assumptions are satisfied, which guarantee the existence of a unique solution of (1).

If b is constant and c and a are linear in x, then equations (1)-(2) reduce to the standard Kalman-filter problem, for which an effective and widely utilised solution exists.

In the nonlinear case, equations describing the evolution of the conditional density were developed by several authors in the mid-1960s (see, for example, Kushner [8]). These can now be found, in their definitive form, in the textbooks of Liptser and Shiryaev [10]. The main result we are refering to ([10], theorem 8.7, pag. 317) states that, under appropriate conditions for a,b,c and x_0 (similar to those of the theorem of existence and uniqueness for the solutions of equations (1) and (2) from [6]), the conditional density of x_t, given $\{y_o, 0 \le s \le t\}$,

$$q_y(t,x) = dP(x_t \le x / y_s, 0 \le s \le t)/dx$$

is the solution of the following nonlinear stochastic parabolic partial differential equation

$$\frac{\partial q_y}{\partial s} = (L_s^y)^* q_y + q_y [c - \int_o^T c(z) q_y(t,z) dz] [dy_t - (\int_o^T c(z) q_y(u,z) dz) du] \quad (3)$$

with initial condition

$$q_y(0,x) = q_o(x)$$

where

$$(L_s^y)^* f = \frac{1}{2} \sum_{i,j} d^{(i,j)}(s,x) \frac{\partial^2 f}{\partial x_i x_j} - \sum_i \frac{\partial}{\partial x_i} [a_y^{(i)}(s,x) f]$$

$$d(t,x) = b(t,x)b^T(t,x)$$

$$a_y^{(i)}(t,x) = a^{(i)}(t,x) - y(t) \sum_j d^{(i,j)}(t,x) \frac{\partial c}{\partial x_j} .$$

Unfortunately, this result has only a theoretical importance, mainly because of the formidable computational problems it entails; for equation (3) no effective solution technique exists. Recently, however, it has been realised that the filtering equations can be recast in substantially simpler form. The main result of the 'pathwise' theory of nonlinear filtering (see Davis [2] and [3]) states that

$$q_y(t,x) = \exp\{y(t)c(x)\}p_y(t,x)/\int_R \exp\{y(t)c(z)\}p_y(t,z)dz$$

where p_y is, under the same conditions as in theorem 8.7 from [10], the solution of the following linear parabolic partial differential equation

$$\begin{aligned} \frac{\partial p_y}{\partial s} &= (L_s^y)^* p_y + e_y p_y \\ p_y(0,x) &= q_0(x) \end{aligned} \tag{4}$$

where

$$e_y(t,x) = \frac{1}{2} y^2(t) \sum_{i,j} d^{(i,j)}(t,x) \frac{\partial c}{\partial x_i} \frac{\partial c}{\partial x_j} - y(t)L_t c(x) - \frac{1}{2}c^2(x)$$

and L_t is the infinitesimal generator of the x_t process. Note that the operator L_s^y can be thought of as the infinitesimal generator of a process z_t satisfying

$$dz_t = a_y(t,z_t)dt + b(t,z_t)dB_t \tag{5}$$

with B_t a standard Wiener process, n-dimensional, with independent components. It follows that equation (4) is the Fokker--Planck equation for z_t, perturbed by a Feynman-Kac transformation, and hence it result that q_y has the probabilistic interpretation:

$$E[f(x_t)/y_s, 0\leq s\leq t] = \frac{E[f(z_t)\exp\{y(t)c(z_t)+\int_0^t e_y(s,z_s)ds\}]}{E[\exp\{y(t)c(z_t)+\int_0^t e_y(x,z_s)ds\}]} \quad (6)$$

This result means that the nonlinear filtering problem is about as simple as it could possible be, in that, to compute the conditional density it is only necessary to solve equations of type (4).

Today, there are a lot of techniques for solving equation (4), however the classical methods are in this case inadequate, because:

- they impose too restrictive conditions for the coefficients of the operator L_s^y;
- they require extensive computational resources in view of the necessity of computing q_y in all the nodes of the discretisation net. In order to avoid these major difficulties, statistical numerical methods of Monte Carlo type are used.

The purpose of the present paper is to develop such a method extending and generalizing the results from [4] to nonlinear filtering problem. The new Monte Carlo method is built-up in section 2. Section 3 gives the algorithm for estimating the conditional expectation of the process x_t using the new method. Finally, section 4 investigates the efficiency of the method from two points of view: the sample size necessary to obtain a given error and the number of operations in order to estimate the solution. A comparision with Kushner method (see for the theoretical aspect [3] and for the computational results [11]) is given.

2. MONTE CARLO SOLUTION OF THE FILTERING EQUATIONS

In the following we consider only deterministic discretizations of the interval $[0,T]$, of $\tau = \{t_i\}_{i=1}^{K}$ type, with $0=t_0<t_1<\ldots<t_K=T$ and $h(\tau) = \max_{i=\overline{1,K}}(t_i-t_{i-1})$ (the random discretization case is outlined in [7]. The new method consists in building-up an approximation z_t^h of the process z_t on which basis, using the unbiased estimator of the arithmetical mean, it's possible to evaluate the conditional expectation of the process x_t given the observations $\{y_s, 0\le s\le t\}$, following the formula

$$E_N^h(x_t/y_s,\ 0\le s\le t) = \frac{\sum_{i=1}^{N} z_k^{h,i} \exp\left\{\sum_{j=0}^{k-1} e_y(t_j,z_j^{k,i})(t_{j+1}-t_j)+y(t)c(z_k^{h,i})\right\}}{\sum_{i=1}^{N} \exp\left\{\sum_{j=0}^{k-1} e_y(t_j,z_j^{k,i})(t_{j+1}-t_j)+y(t)c(z_k^{h,i})\right\}} \tag{7}$$

where $t\in[t_k,t_{k+1})$, $k=\overline{1,K-1}$

$z_j^{h,i} = z_{t_j}^{h,i}$ where the upper index i means the belonging to the i-th pathwise generated from a sample of N trajectories.

We use for z_t^h an iterative approximation of a generalization of Itô formula, given by Platen in [12], satisfying

$z_o^h = z_o$ a sample value $q_o(x)$ distributed

$$z_t^h = z_k^h + \sum_{\alpha\in A_p} f_\alpha(k,z_k^h)I_\alpha(k,t), \quad t\in[t_k,t_{k+1}) \qquad (8)$$

where $p\in N$

$$A_p = \{\alpha\in M/\ell(\alpha)+n(\alpha)\le p, \text{ or } \ell(\alpha) = n(\alpha) = (p+1)/2\}$$

$$f_\alpha = \left\{f_\alpha^{(i)}\right\}_{i=1}^m \qquad f_\alpha^{(i)} = \begin{cases} 0 & \alpha = V \\ a_Y^{(i)} & \alpha = (0) \\ b^{(i,j)} & \alpha = (j),\ j\in\{1,\dots,n\} \\ L^{(j_1)} f_{-\alpha}^{(i)} & \alpha = (j_1,\dots,j_\ell),\ \ell\ge 2 \end{cases}$$

$$I_\alpha(k,t) = \begin{cases} 1 & \alpha = V \\ \int_{t_k}^{t} I_{\alpha-}(i,u)dB_u^{j_\ell}, & \alpha = (j_1,\dots,j_\ell) \end{cases}$$

$$M = \{(j_1,\dots,j_\ell)/j_i\in\{0,1,\dots,n\}\},\ \ell = \overline{1,p}\}\cup\{V\}$$

$$\ell(\alpha) = \text{card}(\alpha),\ n(\alpha) = \text{card}\{j_i=0/j_i \text{ component of } \alpha\}$$

$$\ell(V) = 0 = n(V)$$

if $\alpha = (j_1,\dots,j_\ell)$, $\ell\ge 1$ $\quad -\alpha = (j_2,\dots,j_\ell)$, $\alpha- = (j_1,\dots,j_{\ell-1})$

$$L^{(\ell)}f = \begin{cases} \frac{\partial f}{\partial s} + L_s f & \ell=0 \\ \sum_i b^{(i,\ell)} \frac{\partial f}{\partial x_i} & \ell\in\{1,\dots,n\} \end{cases}$$

$$dB_u^{(0)} = du$$

With this, we give the follosing

<u>Theorem 1.</u> For fixed $p\in N$ and $h>0$, if

(i) $(\forall)\alpha\in A_p$, $s\in[0,T]$, $x,y\in R^m$, $(\exists)c_1$ such that

$\quad ||f_\alpha(s,x)-f_\alpha(s,y)|| \le c_1\ ||x-y||$;

(ii) $(\exists)G$ such that $A_{p-1} \subseteq G \subsetneq A_p$ and $(\forall)\alpha \in G \cup \{\alpha \in M-G / -\alpha \in G\}$

f_α exist and $E\int_0^T ||f_\alpha||^2 ds < \infty$;

(iii) $(\forall)\alpha \in G \cup \{\alpha \in M-G/-\alpha \in G\}$, $s \in [0,T]$, $x \in \mathbb{R}^m$, $(\exists)c_2$

$||f_\alpha(s,x)|| \leq c_2(1+||x||)$;

(iv) $(\forall)\alpha \in (A_p - G) \cup A_1$, $s,t \in [0,T]$, $x \in \mathbb{R}^m$, $(\exists)c_3$

$||f_\alpha(s,x) - f_\alpha(t,x)|| \leq c_3(1+||x||)|s-t|^{1/2}$;

(v) $E||z_0||^2 < \infty$;

(vi) $E||z_0 - z_0^h||^2 \leq c_4 h^p$,

then

$$E \sup_{\substack{t \in [t_k, t_{k+1}) \\ k = \overline{1,K-1}}} ||z_t - z_t^h||^2 \leq c_5 h^p \tag{9}$$

For proof see [12].

For $G=A_p$ and $p=1,2$ we obtain the approximations from [4] and [7]. Obviously, the higher order of convergence for greater p is the result of using more multiple stochastic integrals which contain a higher degree of information on the Wiener process.

Theorem 2. Under the assumptions of theorem 1, $\{z_t^h\}$ is tight in $D^m[0,T]$ (the space of m-dimensional cadlag functions defined on $[0,T]$). If h indexes any convergent sequence with limit z_t, then z_t has continuous paths w.p.1.

For the proof we use relation (9) and the proving technique of theorem 6.3.1 from [9].

Theorem 3. If $f:[0,T]\to\mathbb{R}^m$ belong to $D^m[0,T]$ and is continuous w.p.1 with respect to the measure induced by x_t on $D^m[0,T]$, then

$$E[f(z_t^h)/z_o^h=z_o]\to E[f(z_t)/z_o] \quad \text{as} \quad h\to 0.$$

For the proof see theorems 6.3.2 and 6.3.3 from [9].

It must be noted that $f(z_t) = \exp[\int_o^t e_y(s,z_s)ds+y(t)c(z_t)]$ used in formula (7) does'nt satisfy the continuity assumption of theorem 3 owing to the randomness of the process y(t). The theorem in none the less true because of relation (9) which involves, following theorem 2, the sample paths continuity of the process z_t.

3. THE ALGORITHM

As it results from the last section the kernel of the method is the approximating formula (8). Based on this formula it is easy to derive an algorithm for Generation of Approximate Sample Pathwise solutions of Itô differential equation (5) (GASPI) with respect to any discretization.

We note that (see [4], [7] and [12] too)

- for p=1

$$A_1 = \{v,(0),(1),\dots,(n)\}$$

$$f_v^{(i)} = 0,\ f_{(o)}^{(i)} = a^{(i)},\ f_{(j_1)}^{(i)} = b^{(i,j_1)}$$

$$I_{(o)}(k,t) = \int_{t_k}^{t} du{*}h,\ I_{(j_1)}(k,t)=\int_{t_k}^{t} dB_u^{(j_1)} = Z_1^{(j_1)}\sim\Phi(0,h)$$

$$z_t^{h,(i)}=z_k^{h,(i)}+ha_y^{(i)}(k,z_k^{h,(i)})+\sum_{j_1=1}^{n} b^{(i,j_1)}(k,z_k^{h,(i)})Z_1^{(j_1)} =$$

$$z_o^{(i)}+h\sum_{\ell=1}^{t\in[t_k,t_{k+1})} a_y^{(i)}(\ell,z_\ell^{h,(i)})+\sum_{\ell=1}^{t\in[t_k,t_{k+1})}\sum_{j_1=1}^{n} b^{(i,j_1)}(\ell,z_\ell^{h,(i)})\cdot Z_{1,\ell}^{(j_1)} \qquad (10.a)$$

- for p=2

$$A_2 = A_1\cup\{(j_1,j_2)/j_1,j_2\in\{1,\ldots,n\}\}$$

$$f_{(j_1,j_2)}^{(i)} = \sum_{\ell=1}^{n}\frac{\partial b^{(i,j_2)}}{\partial x_\ell}\, b^{(\ell,j_1)}$$

$$I_{(j_1,j_2)}(k,t) = \int_{t_k}^{t}\int_{t_k}^{r} dB_u^{(j_1)}dB_r^{(j_2)} = (Z_1^{(j_1)}Z_1^{(j_2)}-h)/2$$

$$z_t^{h,(i)} = z_k^{h,(i)}+ha_y^{(i)}(k,z_k^{h,(i)})+\sum_{j_1=1}^{n} b^{(i,j_1)}(k,z_k^{h,(i)})Z_1^{(j_1)} +$$

$$+\sum_{j_1,j_2=1}^{n}\sum_{i_1=1}^{m} b^{(i_1,j_1)}\frac{\partial b^{(i,j_2)}}{\partial x_{i_1}}(Z_1^{(j_1)}Z_1^{(j_2)}-h)/2 =$$

$$= z_o^{(i)}+h\sum_{\ell=1}^{t\in[t_k,t_{k+1})} a_y^{(i)}(\ell,z_\ell^{h,(i)})+\sum_{\ell=1}^{t\in[t_k,t_{k+1})} b^{(i,j_1)}(\ell,z_\ell^{h,(i)})$$

$$\cdot Z_{1,\ell}^{(j_1)}+\sum_{\ell=1}^{t\in[t_k,t_{k+1})}\sum_{j_1,j_2=1}^{n}\sum_{i_1=1}^{m} b^{(i_1,j_1)}\frac{\partial b^{(i,j_2)}}{\partial x_{i_1}}(Z_{1,\ell}^{(j_1)}Z_{1,\ell}^{(j_2)}-h)/2 \qquad (10.b)$$

We stop here with the formulae, because, for $p\geq 3$ the improvement of accuracy does not seem to be so considerable in relation to the more complicated computation apparatus.

Under the conditions of theorem 1 and the assumption of "exact" generators of normal random variables with mean 0 and variance h (see for this point [13]) the algorithm for generation approximate sample pathwise of process z_t is the following

GASPI0. Read data, initialization.

Read as input data :

z_0 - the initial value of the process;

t - the moment at which we have observations about the process;

h - the step of the discretization net;

N_0 - a natural value to initialize the random number generator;

p - the type of the used approximating formula;

n,m - the vectors and matrices z_t^z, Z_1, a and b sizes.

We initialize $Z \leftarrow Z_0$, $TK \leftarrow 0$.

GASPI1. Initial calculus.

We compute k for which $t \in [t_k, t_{k+1})$.

GASPI2. If $TK \geq k$ then go to GASPI5.

GASPI3. Generate a sample value of the random vector $Z_{1,TK}$.

Compute a(TK,Z),b(TK,Z) and, if we use formula (11.b), $\partial b(TK,Z)/\partial x$

GASPI4. $Z \leftarrow Z$ + the right member of (10.a), respectively (10.b), formula according to p=1, or p=2.

$TK \leftarrow TK+1$.

Go to GASPI2.

GASPI5. STOP.∗

A sample pathwise of the process z_t was generated. To compute the conditional expectation, we apply formula (7), using the GASPI algorithm together with N, the sample size and $\{y_s, 0 \leq s \leq t\}$ the noisily observations of process x_t in the points net.

4. THE EFFICIENCY OF THE METHOD

We study the efficiency of the method built-up in section 2, from two points of view:

- the sample size necessary to obtain a given error with a known and big probability;
- the number of operations to estimate the conditional expectation (7) using the new Monte-Carlo method, compared to the Kushner's method from [3].

4.1. The sample size necessary to obtain a given error

Given $\varepsilon>0$ and $h,\delta>0$ small, we estimate N such that

$$P(|E(x_t / y_s, 0\le s\le t) - E_N^h(x_t / y_s, 0\le s\le t)|<\varepsilon)>1-2\delta.$$

Applying a technique used in [1], we obtain for $\delta=0.0455$

$$N = [4\sigma^2/\varepsilon^2] + 1 \tag{11}$$

where we denote by $[\cdot]$ the function 'integer part', and by $\sigma^2 = Var(x_t / y_s, 0\le s\le t)$.

For the evaluation of N, we refer to Glorennec who suggests in [7] to use a χ^2-test on two independent realisations of the process z_t^h.

4.2. The number of operations to estimate the conditional expectation

To estimate the amount of work required to obtain a given level of accuracy, we make the following assumptions:

a) comparing the operations of multiplication-division to the operations of addition-substraction, the last are unsignificant as to the computing time and we will not take them into account when calculating the number of operations necessary to estimate the conditional expectation;

b) the generation of a n-dimensional $\Phi(0,h)$ distributed random vector is comparable, as to the computing time, to the duration of 17n multiplication-division operations. For this estimation we consider that the generation of a $\Phi(0,h)$ distributed random variable, last as 17 multiplications-divisions (we consider here the central-limit-theorem algorithm, i.e. $Z=(U_1+...+U_{16})/16$, where U_i are uniformly i.i.d. variables, see [13] too), and the algorithm for the random vector generation, consist, in our case (see [13] too), in generation of n normlay i.i.d. variables, because the covariance matrix are diagonal.

c) the evaluation of a_y and b in a sample pathwise point is comparable, as to the computing time, to the duration of m, respectively mn, multiplication-division operations because, we presume, that the one component estimation of the vector a_y, respectively of the matrix b last as a multiplication-division. Though not generally true in practice, this assumption is made only to compare our algorithm with Davis' one.

Thus, the number of operations necessary to generate a z_t^h-sample pathwise of length k, is

$$N_{Z,1} = k(2mn+2m+17n) \qquad \text{for formula (10.a)}$$

and

$$N_{Z,2} = k(m^2n^2+m^2n+2mn+2m+17n) \qquad \text{for formula (10.b)}$$

then, to estimate the conditional expectation, we need

$$N_{E,i} = N(15 + 3k + N_{Z,i}) \quad i=1,2 \text{ operations.} \tag{12}$$

Under the same assumptions, using the techniques from [5] and [9], we obtain for Kushner's method from [3] the following results:

$$E \sup_{\substack{t\in[t_k,t_{k+1})\\ k=\overline{1,K-1}}} ||z_t-z_t^h|| \leq Ch/(\Delta^2 Q)$$

and

$$N_Z = k(5 + 2m + 2mn) + 2.$$

The formula is to establish for the algorithm presented in [3] pag. 171. Following Davis'notations and ours hypothesis, we consider: two operations for the generation of $\xi\delta_o^\Delta$, and for the multiplication $f(x_i)d_i$; and for each step of the pathwise of length k: five operations for the generation of U, evaluations $\delta\tilde{c}$ and exp(sum+yh(x)), and for the selection of a new state; 2m+2mn operations for the evaluations of c^y and b^y and a(t,x) from the transition probabilities $\{p_t\}$.

Comparing this result, we note that

- for the same level of accuracy our method is more rapidly than Kushner'method;
- the new algorithm allows, when necessary, to increase the convergence speed, compared with Kushner-algorithm which does not allow such possibility. On the other hand, it's natural that, the improvement of the convergence speed to imply the growth of the number of operations. Consequently, we claim that, for its simplicity, generality and efficiency, our algorithm is more convenient than Kushner's algorithm.

REFERENCES

[1] CURTISS, J.H. - "A theoretical comparision of the efficiencies of two classical methods and a Monte Carlo method for computing one component of the solution of a set of linear algebraic equations", 1956, Symp. on Monte Carlo methods, ed. H.A. Meyer, Wiley, N.Y.

[2] DAVIS, M.H.A., WELLINGS, P.H. - "Computational problems in nonlinear filtering", 1980, Analysis on Optimization of Systems', Lecture notes in control and information sciences, 28, Springer-Verlag, Berlin

[3] DAVIS, M.H.A. - "New approach to filtering for nonlinear systems" 1981, IEEE Proc. D, Control Theory & Appl., 128, 5, pp. 166-172.

[4] ENACHESCU, D.M. - "Monte Carlo methods for solving a class of partial differential equations with variable coefficients", 1975, Bull, Math., 19(67), 3-4, pp. 277-282

[5] Ibid. - "Studiul satistic al proceselor stocastice cu aplicatii la ecuatiile parabolice (Statistic study of the stochastic processes, computing aspects to the parabolic equation) in romanian, 1980, Doctorate Thesis, Univ. of Bucharest

[6] GIHMAN, I.I., SKOROHOD, A.V. - "Stochastic differential equations" 1972, Springer-Verlag, Berlin

[7] GLORENNEC, P.Y. - "Estimation a priori des erreurs dans la résolution numérique d'équations differentielles stochastiques", 1977, Séminaire de probabilités, Univ. de Rennes, fasc. 1, pp. 57-93.

[8] KUSHNER, H.J. - "On the differential equations satisfied by conditional probability densities of Markov processes", 1964, SIAM J. Contr., 2, pp. 106-119

[9] Ibid. - "Probability methods for approximations in stochastic control and for elliptic equations", 1977, Academic Press, N.Y.

[10] LIPTSER, R.S., SHIRAYAEV, A.N. - "Statistics of random processes I", 1977, Springer-Verlag, New York

[11] O'LOGHLER, J.W., WRIGHT, G.C. - "Computer results for new nonlinear filtering algorithm", 1982, IEE Proc. D, Control Theory & Appl., 129, 2, pp. 70-71

[12] PLATEN, E. - "An approximation method for a class of Itô processes", 1979, Preprint ZIMM, AdW der DDR, Berlin

[13] Văduva, I. - "RAVAGE. A subroutine library for computer generation of random numbers, random variables, random vectors and stochastic processes", 1976, GMD--Mitteilungen, 9, Bonn

Denis M. Enăchescu
University of Bucharest
Computing Centre
Str. Stefan Fortună no. 125
R-77116 Bucharest

Proc. of the 4th Pannonian Symp. on Math. Stat.. Bad Tatzmannsdorf, Austria 1983
W. Grossmann, G. Pflug, I. Vincze, W. Wertz, eds

LOCAL PROPERTIES OF FUNCTION FITTING ESTIMATES WITH APPLICATION TO SYSTEM IDENTIFICATION

ALEXANDER A. GEORGIEV

Wrocław

ABSTRACT

This paper presents a general family of nonparametric estimates of the unknown function with nonrandom domain. Sufficient conditions for weak and strong pointwise convergence of estimates are derived. A sharper result for the Priestley--Chao kernel estimate is shown. A new consistent sequence of estimates, defined in terms of the k-nearest neighbour (K-NN), is constructed. These results are applied to the problem of identifying noisy linear dynamic systems.

1. INTRODUCTION

Let $(x_1, Y_1), \ldots, (x_n, Y_n)$ be independent pairs of observations described by the following nonlinear model

$$E(Y_i) = g(x_i), \quad i=1,2,\ldots,n, \tag{1}$$

where $\{x_1,\ldots,x_n\}$ are fixed regressor variables in $[0,1]$. The problem in this paper is to estimate the unknown function g, provided that it is "smooth". We shall consider a class of nonparametric estimates of g. The "nonparametric" property refers to the absence of finite dimensional continuous parametrization of the space of functions containing g. Estimation of a completely unknown function was first considered by Priestley and Chao [7] and later studied by Mustafi [6], Clark [4], Benedetti [1], Gasser and Müller [5], Schuster and Yakowitz [11], Cheng and

Lin [2], [3], as well as Rutkowski [9], [10].

Any Borel measurable function of x and the data will be called a function fitting estimate. Since the function g is local by nature, the function fitting estimate can be regarded as an appropriate average of observations in the neighborhood of the point under consideration. We estimate g(x) in (1) using the following formula

$$g_n(x) = \sum_{i=1}^{n} w_{ni}(x)\, Y_i, \tag{2}$$

where $(w_{n1}(x), w_{n2}(x), \ldots, w_{nn}(x))$ is a vector of weights, $w_{ni}(x) = w_{ni}(x;\, x_1, x_2, \ldots, x_n)$, $i=1,2,\ldots,n$. Weights of values of i for which x_i is close to x are heavier than those for which x_i is far from x. Set $w_{ni}(x) = 0$ for $i>n$.

On the basis of the research literature on the topic at hand, we believe that this is the first study of general non-parametric function fitting, when the domain points are non-random. We shall investigate pointwise weak and almost sure behaviour of g_n (Section 2) and show sharper results for the Priestley-Chao estimate with kernel weights (Section 3). Further, we shall introduce a new sequence of k-nearest neighbor (k-NN) weights (Section 4) and apply the nonparametric function fitting methodology to the problem of identification of linear dynamic systems (Section 5).

2. THE MAIN RESULT

We state the following theorem which establishes the

pointwise convergence of the estimate g_n. Let us define

$$U_i = Y_i - E(Y_i), \quad i=1,2,\dots,n. \tag{3}$$

Theorem 1. Let $\{w_{ni}(x)\}$ be a sequence of weights and let the following conditions be satisfied: there is $b>0$ such that

$$\sum_{i=1}^{n} |w_{ni}(x)| \le b \qquad \text{for all } n\ge 1 , \tag{4}$$

$$\sum_{i=1}^{n} w_{ni}(x) \to 1 , \qquad \text{as } n\to\infty \tag{5}$$

and

$$\sum_{i=1}^{n} |w_{ni}(x)| I_{\{|x_i-x|>a\}} \to 0, \text{ as } n\to\infty \text{ for all } a>0. \tag{6}$$

If, in addition, we assume that $\{U_i\}$ is a sequence of i.i.d. random variables and

$$\left.\begin{array}{l} E|U_1| < \infty \\ \sup_i |w_{ni}(x)| \to 0, \text{ as } n\to\infty, \end{array}\right. \tag{7^a}$$

or

$$\left.\begin{array}{l} E|U_1|^{1+1/s} < \infty , \\ \sup_i |w_{ni}(x)| = 0(n^{-s}) , \end{array}\right. \tag{7^b}$$

then $g_n(x)\to g(x)$ in probability or with prob. 1 respectively (8) at every continuity point x of g.

Proof. Choose $a>0$ and x as a continuity point of g. First note that

$$|g_n(x)-g(x)| \le |g_n(x)-Eg_n(x)| + |Eg_n(x)-g(x)|. \tag{9}$$

For the bias term of (9) we have

$$|Eg_n(x) - g(x)| \le$$

$$\le \sum_{i=1}^{n} |w_{ni}(x)| \cdot |g(x_i) - g(x)| I_{\{|x_i - x| \le a\}} +$$

$$+ \sum_{i=1}^{n} |w_{ni}(x)| \cdot |g(x_i) - g(x)| I_{\{|x_i - x| > a\}} +$$

$$+ |g(x)| \cdot \left| \sum_{i=1}^{n} w_{ni}(x) - 1 \right| \le$$

$$\le \varepsilon \cdot b + 2c \sum_{i=1}^{n} |w_{ni}(x)| I_{\{|x_i - x| > a\}} + c \left| \sum_{i=1}^{n} w_{ni}(x) - 1 \right| \qquad (10)$$

where $c = \sup_{x \in [0,1]} |g(x)|$ and $\varepsilon = \sup_{|x-y| \le a} |g(x) - g(y)|$. (10) tends to zero if we let first $n \to \infty$ and then $a \to 0$.

Now, we need only to show that $|g_n(x) - Eg_n(x)| \to 0$ as $n \to \infty$, in probability (with prob. 1). Obviously

$$|g_n(x) - Eg_n(x)| = | \sum_{i=1}^{n} w_{ni}(x) U_i | . \qquad (11)$$

By virtue of Pruitt's [8] Theorem 1 and 2, the proof of the Theorem is complete under our assumptions (7).

3. KERNEL WEIGHTS

An estimate

$$\hat{g}_n(x) = \sum_{i=1}^{n} \hat{w}_{ni}(x) Y_i \qquad (12)$$

with kernel weights

$$\hat{w}_{ni}(x) = K\left(\frac{x - x_i}{h_n}\right) \frac{x_i - x_{i-1}}{h_n} \qquad (13)$$

was first considered by Priestley and Chao [7], where K is a bounded function (so-called kernel) defined over R^1, h_n is a nonnegative sequence of numbers decreasing to zero as $n \to \infty$, and the design variables $\{x_i\}$ are assumed to be known without error and satisfy the following order condition

$$0 = x_0 < x_1 < \ldots < x_n \leq 1. \tag{14}$$

Other kernel weights have been studied by Clark [4], Gasser and Müller [5] and Cheng and Lin [2], [3].

In view of Theorem 1 we have

<u>Theorem 2</u>. Let us suppose that K is a continuous probability density function, such that K(x) is nonincreasing for x>0, and nondecreasing for x<0. Let

$$\Delta_n = \max_{1 \leq i \leq n} |x_i - x_{i-1}| = O(n^{-1}) . \tag{15}$$

If $\{U_i\}$ is a sequence of i.i.d. random variables,

$$h_n \to 0, \text{ as } n \to \infty , \tag{16}$$

and

$$E|U_1| < \infty , \quad nh_n \to \infty, \text{ as } n \to \infty , \tag{17^a}$$

or

$$E|U_1|^{1+1/s} < \infty , \quad h_n^{-1} = O(n^{1-s}) \text{ for some } 0<s<1 \tag{17^b}$$

then $\hat{g}_n(x) \to g(x)$ in probability or with prob. 1 resp. (18) at every continuity point $x \in (0,1)$ of g.

Proof. Since $\{\hat{w}_{ni}(x)\}$ is a sequence of nonnegative weights ($\hat{w}_{ni}(x)\geq 0$), condition (4) is implied by (5). Let $x\in(0,1)$. First, we check condition (5).

$$\sum_{i=1}^{n} \hat{w}_{ni}(x) = \sum_{i=1}^{n} K\left(\frac{x-x_i}{h_n}\right) \frac{x_i-x_{i-1}}{h_n} \to 1 \text{ , as } n\to\infty \tag{19}$$

from Lemma 1 in Benedetti [1], p. 250, and by the assumptions of Theorem 2. Consider (6) now:

$$\sum_{i=1}^{n} |\hat{w}_{ni}(x)| I_{\{|x_i-x|>a\}} = \sum_{i=1}^{n} K\left(\frac{x-x_i}{h_n}\right) \frac{|x_i-x_{i-1}|}{h_n} I_{\{|x_i-x|>a\}} \leq$$

$$\leq \frac{n\cdot\Delta_n}{a} \sup_{|y|>\frac{a}{h_n}} |y|K(y)\to 0, \text{ as } n\to\infty \text{ ,} \tag{20}$$

if our conditions for $K(\cdot)$ and (15), (16) are satisfied. We now prove that (7) holds if (17) is satisfied, i.e.

$$\sup_i |\hat{w}_{ni}(x)| \leq \frac{\sup K(y)}{nh_n} \to 0, \text{ as } n\to\infty \tag{21^a}$$

and

$$n\cdot\sup_i |\hat{w}_{ni}(x)| \leq h_n^{-1} \sup K(y) = 0(n^{1-s}), \text{ resp.} \tag{21^b}$$

Remark 1. The conditions for the weak version of the above Theorem are weaker than those imposed by Priestley and Chao [7], Theorem 1, i.e.

$$g(x)\in \mathrm{Lip}(\alpha),\ K(x)\in \mathrm{Lip}(\beta),\ nh_n^{p}\to\infty,\ p\geq 2, \tag{22}$$

and by Cheng and Lin [2], Theorem 1(iii), i.e.

$$g(x) \text{ is continuous in } [0,1],\ K(x)\in \mathrm{Lip}(\beta),\ nh_n^{1+1/\beta}\to\infty\ . \tag{23}$$

Remark 2. The conditions for the strong version are more constructive from the practical point of view and better than those obtained by Benedetti [1], Theorem 3,

$$g(x) \in \mathrm{Lip}(\alpha),\ K(x) \in \mathrm{Lip}(\beta),\ E|U_1|^4 < \infty,\ \sum_{n=1}^{\infty} \left(\frac{1}{nh_n^2}\right)^2 < \infty . \tag{24}$$

4. NEAREST NEIGHBOR WEIGHTS

In this section, a different sequence of weights will be constructed. The disadvantage of estimates with kernel weights is that the degree of smoothing around the observation points is chosen without any regard to the data. On the basis of the data sequence, we construct a new estimate

$$\check{g}_n(x) = \sum_{i=1}^{n} \check{w}_{ni}(x)\, Y_i , \tag{25}$$

with the nearest neighbor weights

$$\check{w}_{ni}(x) = K\left(\frac{x-x_i}{r_n}\right) \frac{x_i-x_{i-1}}{r_n} , \tag{26}$$

where K is a bounded function, $\{x_i\}$ satisfy (14), and $r_n = r_n(x, k_n, x_1, \ldots, x_n)$ is the distance between x and the k_n-th nearest neighbor (k - NN) observation of x among all x_i's. The number of neighbors k_n depends on n and is determined by the experimenter.

The following Theorem establishes sufficient conditions for the consistency of the estimate $\check{g}_n$.

Theorem 3. Let us suppose that K is a continuous probability density function, such that K(x) is nonincreasing for x>0, and nondecreasing for x<0. Let (15) be satisfied and

$$\delta_n = \min_{1\le i\le n} |x_i - x_{i-1}| \ge d/n, \ d>0 \ . \tag{27}$$

If $\{U_i\}$ is a sequence of i.i.d. random variables,

$$k_n/n \to 0, \text{ as } n\to\infty, \tag{28}$$

and

$$\left.\begin{aligned} &E|U_1| < \infty, \\ &k_n \to \infty, \text{ as } n \to \infty, \end{aligned}\right. \tag{29^a}$$

or

$$\left.\begin{aligned} &E|U_1|^{1+1/s} < \infty , \\ &k_n^{-1} = O(n^{-s}) \text{ for some } 0<s<1, \end{aligned}\right. \tag{29^b}$$

then $\check{g}_n(x)\to g(x)$ in probability or with prob. 1 respectively (30) at every continuity point $x\in(0,1)$ of g.

<u>Proof</u>. Note that

$$k_n\delta_n \le r_n \le k_n\Delta_n \tag{31}$$

by the definition of r_n. The proof is similar to that in Theorem 2.

We define estimate (2) as an appropriate average of the observations in the neighborhood of the point x. In the kernel method a region S, based on h_n, is formed around x without regard to the data. The nearest neighbor method allows the data to play a role in the degree of smoothing. The smallest sphere Q, containing the k_n nearest neighbors of x among the observations, is found. In each method, the estimates are obtained after weight-averaging of Y_i of the appropriate region

(S in the kernel case, Q in the k-NN case). The correspondence $nh_n \leftrightarrow k_n$ seems to hold as in the regression estimation for the model $m(x) = E(Y|X=x)$. Both estimates have the same asymptotic properties.

5. APPLICATION TO SYSTEM IDENTIFICATION

Procedure (2), with suitable weights (13) or (26), can be used for the identification of the frequency responce of a linear dynamic system. Let us assume that $H(j\omega)$ is a transfer function and the input signal is $u(t) = \sin(\omega t)$. It is known that the output signal is

$$y(t) = a(\omega) \sin(\omega t + \varphi(\omega)) , \tag{32}$$

where $a(\omega) = |H(j\omega)|$ and $\varphi(\omega) = \arg H(j\omega)$. Let us suppose that 2n noisy measurements are to be made in $\{\omega_1,...,\omega_n\}$, i.e.

$$A_i = |H(j\omega_i)| + Z_{1i}, \quad i=1,2,...,n, \tag{33}$$

and

$$\Phi_i = \arg H(j\omega_i) + Z_{2i}, \quad i=1,2,...,n. \tag{34}$$

The noisy measurements $\{A_i\}$ and $\{\Phi_i\}$ and frequency points $\{\omega_i\}$ are enough for a consistent estimation of $a(\omega)$ and $\varphi(\omega)$ by procedure (2).

6. FINAL REMARKS

A general family of nonparametric estimates for the function fitting is proposed. Conditions for the pointwise convergence of the above procedure are derived. New results for the known kernel estimate and for the new k-nearest neighbor estimate are given. Unfortunately, this family does not include the Rutkowski's orthogonal series estimate, because the weights - based on orthogonal series - do not form the Teoplitz summation matrix, i.e. the conditions (4), (5), (7^a) (see [8]) are not satisfied.

Further work may be concerned with the general estimation of the derivative of g(x).

REFERENCES

[1] BENEDETTI, J. (1977). On the nonparametric estimation of regression function, J. Roy. Statist. Soc. B, 39, 248-253.

[2] CHENG, K.F. and LIN, P.E. (1981). Nonparametric estimation of regression function, Z. Wahrscheinlichkeitstheorie verw. Gebiete, 57, 223-233.

[3] CHENG, K.F. and LIN, P.E. (1981). Nonparametric estimation of a regression function: limiting distribution, Australian J. Statist., 23, 186-195.

[4] CLARK, R.M. (1977). Nonparametric estimation of a smooth regression function, J. Roy. Statist. Soc. B, 39, 107-113.

[5] GASSER, T. and MÜLLER, H.G. (1979). Kernel estimation of regression function. In Smoothing Techniques for Curve Estimation, LNM 757, Gasser, T. and Rosenblatt, M. Eds., Springer, Berlin-New York, 23-68.

[6] MUSTAFI, C.K. (1976). A note on non-parametric curve fitting, Utilitas Math., 10, 129-137.

[7] PRIESTLEY, M.B. and CHAO, M.T. (1972). Nonparametric function fitting, J. Roy. Statist. Soc. B, 34, 385-392.

[8] PRUITT, W.E. (1966). Summability of independent random variables. J. Math. Mech., 15, 769-776.

[9] RUTKOWSKI, L. (1982). On system identification by non-parametric function fitting, IEEE Trans. Autom. Control, 27, 225-227.

[10] RUTKOWSKI, L. (1982). Orthogonal series estimate of a regression function with applications in system identification, In Probability and Statistical Inference, Grossmann, W., Pflug, G. Ch. and Wertz, W. Eds., D. Reidel, Dordrecht Holland, 343-347.

[11] SCHUSTER, E. and YAKOWITZ, S. (1979). Contribution to the theory of nonparametric regression with application to system identification, Ann. Statist., 7, 139-149.

Alexander A. GEORGIEV
Technical University
Institute of Kybernetics
St. Wyspiańskiego 27
PL-50-370. WROCŁAW

Proc. of the 4th Pannonian Symp. on Math. Stat., Bad Tatzmannsdorf, Austria 1983
W. Grossmann, G. Pflug, I. Vincze, W. Wertz, eds

BAYESIAN ESTIMATION OF MEANS IN q-COMPONENT HIERARCHICAL DESIGN WITH RANDOM EFFECT MODEL

HELENA JELENKOWSKA

Lublin

Summary

The problem of estimating the means in a q-component hierarchical random effect model

$$y_{i_q \ldots i_1} = \mu + \alpha_{i_q} + \beta_{i_q i_{q-1}} + \ldots + \delta_{i_q \ldots i_2} + e_{i_q \ldots i_1}$$

$$(i_q = 1 \ldots I_q;\ i_1 = 1 \ldots I_1)$$

is considered from a Bayesian point of view. Posterior distributions of α_{i_q}, $\beta_{i_q i_{q-1}}, \ldots, \delta_{i_q \ldots i_2}$ are obtainded under the assumptions that these effects are normally and independently distributed with zero means and variances $\sigma_q^2, \sigma_{q-1}^2, \ldots, \sigma_2^2$ respectively and the $e_{i_q \ldots i_1}$ are independent random errors from $N(0, \sigma_1^2)$ and adopting a non-informative reference prior distribution for $\mu, \sigma_1^2, \ldots, \sigma_q^2$.

1. Introduction

Tiao and Box (1968) have studied the problem of Bayesian estimation of the means in the one-way model, Giri (1970) for the two-way classification without interaction, Clement (1975) for the two-way classification with interaction and Clement, Giri (1977) for the three - component hierarchical design model.

We consider here the q-component hierarchical model

$$(1.1)\quad y_{i_q \ldots i_1} = \mu + \alpha_{i_q} + \beta_{i_q i_{q-1}} + \ldots + \delta_{i_q \ldots i_2} + e_{i_q \ldots i_1}$$

$$(i_q = 1 \ldots I_q, \ldots, i_1 = 1 \ldots I_1)$$

where $y_{i_q \ldots i_1}$ are the observations, μ is a common location paraméter α_{i_q}, $\beta_{i_q i_{q-1}}, \ldots, \delta_{i_q \ldots i_2}$ are q-1 different kinds of effects and $e_{i_q \ldots i_1}$ are random errors or disturbances. Throughout this article we will write $N(\mu, \sigma^2)$ for the normal distribution with mean μ and variance σ^2 and we will reserve the symbol "~" for the term "distributed as".

In random effect model, the usual assumptions are that the random effects $\alpha_{i_q}, \beta_{i_q i_{q-1}}, \ldots, \delta_{i_q \ldots i_2}$ and the errors $e_{i_q \ldots i_1}$ are all indepedent and

$$(1.2)\quad \alpha_{i_q} \sim N(0, \sigma_q^2),\ \beta_{i_q i_{q-1}} \sim N(0, \sigma_{q-1}^2), \ldots, \delta_{i_q \ldots i_2} \sim N(0, \sigma_2^2),$$

$$e_{i_q \ldots i_1} \sim N(0, \sigma_1^2).$$

2. Posterior distribution of the vector

$$\left(\frac{\sigma_1^2}{\sigma_{12}^2}, \frac{\sigma_{12}^2}{\sigma_{123}^2}, \ldots, \frac{\sigma_{12\ldots q-1}^2}{\sigma_{12\ldots q}^2}\right).$$

We assume the non-informative reference prior for $(\mu, \sigma_1^2, \sigma_{12}^2, \ldots, \sigma_{12\ldots q}^2)$ as

$$(2.1) \quad f(\mu, \sigma_1^2, \sigma_{12}^2, \ldots, \sigma_{12\ldots q}^2) = c\sigma_1^{-2}, \sigma_{12}^{-2} \ldots \sigma_{12\ldots q}^{-2}$$

where

$$\sigma_{12}^2 = \sigma_1^2 + I_1\sigma_2^2$$

$$\sigma_{123}^2 = \sigma_1^2 + I_1\sigma_2^2 + I_1I_2\sigma_3^2$$

$$\vdots$$

$$\sigma_{12\ldots q}^2 = \sigma_1^2 + I_1\sigma_2^2 + \ldots + I_1I_2\ldots I_{q-1}\sigma_q^2$$

and μ, $\log \sigma_1^2, \ldots, \log \sigma_{12\ldots q}^2$ are locally independent and uniform.

Tiao and Box (1967) have derived the posterior distribution of $(\sigma_1^2, \ldots, \sigma_{12\ldots q}^2)$, as

$$(2.2) \quad f(\sigma_1^2, \sigma_{12}^2, \ldots, \sigma_{12\ldots q}^2 | y) = \frac{1}{\gamma} f_{i\chi^2}\left(\frac{\sigma_1^2}{\nu_1 m_1} \middle| \nu_1\right) f_{i\chi^2}\left(\frac{\sigma_{12}^2}{\nu_2 m_2} \middle| \nu_2\right)$$

$$\ldots f_{i\chi^2}\left(\frac{\sigma_{12\ldots q}^2}{\nu_q m_q} \middle| \nu_q\right),$$

where $m_1, \ldots, m_q$ are mean squares, $\nu_1, \ldots, \nu_q$ are degree of freedom and

$$\gamma = \prod_{i=1}^{q} (\nu_i m_i)^{-1} P_{q-1}\left(\frac{\nu_2 m_2}{\nu_1 m_1}, \frac{\nu_3 m_3}{\nu_2 m_2}, \ldots, \frac{\nu_q m_q}{\nu_1 m_1}; \frac{\nu_2}{2}, \frac{\nu_3}{2}, \ldots, \frac{\nu_q}{2}; \frac{\nu_1}{2}\right)$$

$f_{i\chi^2}$ is the inverted chisquare density function and P_{q-1} is the cumulative probability function of an inverted (q-1)-dimensional Dirichlet distribution.

The posterior distribution of

$$(2.3) \quad x_1 = \frac{\sigma_1^2}{\sigma_{12}^2}, \qquad x_2 = \frac{\sigma_1^2}{\sigma_{123}^2}, \ldots, x_{q-1} = \frac{\sigma_{12\ldots q-1}^2}{\sigma_{12\ldots q}^2}$$

can be obtained from (2.2) by making a transformation from

$$(\sigma_1^2, \sigma_{12}^2, \ldots, \sigma_{12\ldots q}^2) \quad \text{to} \quad \left(\frac{\sigma_1^2}{\sigma_{12}^2}, \frac{\sigma_{12}^2}{\sigma_{123}^2}, \ldots, \frac{\sigma_{12\ldots q-1}^2}{\sigma_{12\ldots q}^2}\right)$$

and integrating out σ_1^2, as

$$(2.4) \quad f(x_1, \ldots, x_{q-1} | y) = \frac{1}{c} \frac{\nu_q m_q}{\nu_1 m_1} \frac{\Gamma(\frac{1}{2} \sum_{m=2}^{q} \nu_m)}{\prod_{m=1}^{q} \frac{\nu_m}{2}} \left(\frac{x_1 \nu_2 m_2}{\nu_1 m_1}\right)^{\frac{\nu_2}{2}} \cdot$$

$$\cdot \left(\frac{x_1 x_2 \nu_3 m_3}{\nu_1 m_1}\right)^{\frac{\nu_3}{2}} \cdots \left(\frac{x_1 x_2 \ldots x_{q-1} \nu_q m_q}{\nu_1 m_1}\right)^{\frac{\nu_q}{2} - 1} \left(1 + \frac{x_1 \nu_2 m_2}{\nu_1 m_1} + \ldots\right.$$

$$\left. \ldots + \frac{x_1 x_2 \ldots x_{q-1} \nu_q m_q}{\nu_1 m_1}\right)^{-\frac{1}{2} \sum_{m=1}^{q} \nu_m}$$

where

$$c = P_{q-1}\left(\frac{\nu_2 m_2}{\nu_1 m_1}, \ldots, \frac{\nu_q m_q}{\nu_1 m_1}; \frac{\nu_2}{2}, \ldots, \frac{\nu_q}{2}; \frac{\nu_1}{2}\right).$$

Thus

$$w_1 = x_1 \frac{\nu_2 m_2}{\nu_1 m_1}, \qquad 0 < w_1 < \frac{\nu_2 m_2}{\nu_1 m_1},$$

$$w_2 = x_1 x_2 \frac{\nu_3 m_3}{\nu_1 m_1}, \qquad 0 < w_2 < \frac{\nu_3 m_3}{\nu_1 m_1},$$

$$\vdots$$

$$w_{q-1} = x_1 x_2 \ldots x_{q-1} \frac{\nu_q m_q}{\nu_1 m_1}, \quad 0 < w_{q-1} < \frac{\nu_q m_q}{\nu_1 m_1}$$

have a joint inverted Dirichlet distribution. Using (2.4) it follows that the noncentral mixed moments of the variables $x_1, x_2, \ldots, x_{q-1}$ are

$$(2.5) \quad \mu_{k_1 \ldots k_{q-1}} = E(x_1^{k_1} \ldots x_{q-1}^{k_{q-1}} | y) = \left(\frac{\nu_1 m_1}{\nu_2 m_2}\right)^{k_1} \left(\frac{\nu_2 m_2}{\nu_3 m_3}\right)^{k_2} \cdots \left(\frac{\nu_{q-1} m_{q-1}}{\nu_q m_q}\right)^{k_{q-1}}$$

$$\frac{\Gamma(\frac{\nu_1}{2} - k_1)\Gamma(\frac{\nu_2}{2} + k_1 - k_2)\ldots\Gamma(\frac{\nu_{q-1}}{2} + k_{q-2} - k_{q-1})\Gamma(\frac{\nu_q}{2} + k_{q-1})}{\prod_{m=1}^{q} \Gamma(\frac{\nu_m}{2})} \cdot \frac{P_{q-1}^{k_1 \ldots k_q}}{P_{q-1}^{0 \ldots 0}}$$

where

$$(2.6)\quad P_{q-1}^{k_1\ldots k_{q-1}} = P_{q-1}\left(\frac{\nu_2 m_2}{\nu_1 m_1},\ldots,\frac{\nu_q m_q}{\nu_1 m_1};\ \frac{\nu_2}{2}+k_1-k_2,\ldots,\frac{\nu_q}{2}+k_{q-1};\frac{\nu_1}{2}-k_1\right)$$

3. Posterior distribution of means in the random effect model

Let

$$a_{i_q} = \alpha_{i_q} - \bar{\alpha}, \qquad \bar{\alpha} = I_q^{-1} \sum_{i_q=1}^{I_q} \alpha_{i_q}$$

$$(3.1)\qquad b_{i_q i_{q-1}} = \beta_{i_q i_{q-1}} - \beta_{i_q}, \quad \beta_{i_q} = I_{q-1}^{-1} \sum_{i_{q-1}=1}^{I_{q-1}} \beta_{i_q i_{q-1}}$$

$$\vdots$$

$$d_{i_q\ldots i_2} = \delta_{i_q\ldots i_2} - \delta_{i_q\ldots i_3}, \quad \delta_{i_q\ldots i_3} = I_2^{-2} \sum_{i_2=1}^{I_2} \delta_{i_q\ldots i_2}$$

$$\underset{\sim}{a} = (a_1,\ldots,a_{I_q}),\ \underset{\sim}{b} = (b_{11},\ldots,b_{I_q I_{q-1}}) \ldots \underset{\sim}{d} = (d_{11\ldots1},\ldots,d_{I_q\ldots I_2})$$

Using (1.1), (1.2) and (2.1) the posterior distribution of $(\mu,\ \underset{\sim}{\alpha},\ \underset{\sim}{\beta},\ldots,\ \underset{\sim}{\delta},\sigma_1^2,\ldots,\sigma_q^2)$ given the vector of observations $\underset{\sim}{y}$ can be derived as

$$(3.2)\quad f(\mu,\underset{\sim}{\alpha},\underset{\sim}{\beta},\ldots,\underset{\sim}{\delta},\sigma_1^2,\sigma_2^2,\ldots,\sigma_q^2|\underset{\sim}{y}) = c\cdot(\sigma_1^2)^{-\frac{1}{2}(\sum_{m=1}^{q}\nu_m+3)}\cdot(\sigma_2^2)^{-\frac{1}{2}(\sum_{m=2}^{q}\nu_m+1)}\ldots$$

$$\ldots\ \sigma_q^{2\,-\frac{1}{2}(\nu_q+1)}\,(\sigma_{12}^2\sigma_{123}^2\ldots\sigma_{12\ldots q}^2)^{-1}\exp\left(-\frac{1}{2\sigma_1^2}A\right),$$

where

$$(3.3)\quad A = \sum_{i_q=1}^{I_q} \cdots \sum_{i_1=1}^{I_1} (y_{i_q\ldots i_1} - \mu - \alpha_{i_q} - \beta_{i_q i_{q-1}} \cdots - \delta_{i_q\ldots i_2})^2$$

$$+ \frac{\sigma_1^2}{\sigma_q^2} \sum_{i_q=1}^{I_q} \alpha_{i_q}^2 + \frac{\sigma_1^2}{\sigma_{q-1}^2} \sum_{i_q=1}^{I_q} \sum_{i_{q-1}=1}^{I_{q-1}} \beta_{i_q i_{q-1}}^2 + \ldots + \frac{\sigma_1^2}{\sigma_2^2} \sum_{i_q=1}^{I_q} \cdots$$

$$\cdots \sum_{i_2=1}^{I_2} \sigma_{i_q\ldots i_2}^2 .$$

We write

$$y_{i_q\ldots i_1} - \mu - \alpha_{i_q} - \beta_{i_q i_{q-1}} - \ldots - \delta_{i_q\ldots i_2} = -[\delta_{i_q\ldots i_2} - (y_{i_q i_{q-1}\ldots}$$

$$\ldots{}_{i_2 \bullet} - y_{i_q\ldots i_3 \bullet\bullet})] - \ldots - [b_{i_q i_{q-1}} - (y_{i_q i_{q-1} \bullet \ldots \bullet} - y_{i_q \bullet \ldots \bullet})] -$$

$$- [\alpha_{i_q} - (y_{i_q \bullet \ldots \bullet} - y_{\bullet \ldots \bullet})] - (\mu + \bar{\alpha} + \bar{\beta} + \ldots + \bar{\delta} - y_{\bullet \ldots \bullet}).$$

Squaring both sides, summing over i_q, i_{q-1}, ..., i_1, using transformation (2.3) to express A, transforming from vector $(\mu, \alpha_{i_q}, \ldots, \delta_{i_q\ldots i_2}, \sigma_1^2, \ldots, \sigma_q^2)$ to vector $(\mu, \bar{\alpha} \ldots \bar{\delta}, \underset{\sim}{a}, \ldots$ $\ldots, \underset{\sim}{d}, \sigma_1^2, x_1, \ldots x_{q-1})$ and integrating out $\bar{\alpha}, \ldots, \bar{\delta}$, one obtains

$$(3.4)\quad f(\underset{\sim}{a}, \ldots, \underset{\sim}{d} \mid \sigma_1^2, x_1, \ldots, x_{q-1}, y) = cH(x_1, \ldots, x_{q-1})(\sigma_1^2)^{-(s+1)}$$

$$\exp\left(- \frac{1}{2\sigma_1^2} B \right),$$

where

$$(3.5) \qquad B = \nu_1 m_1 + x_1 \nu_2 m_2 + \ldots + x_1 \ldots x_{q-1} \nu_q m_q +$$

$$+ \frac{I_1}{1+x_1} \sum_{i_q} \ldots \sum_{i_2} \left[d_{i_q \ldots i_2} - (1-x_1)(y_{i_q \ldots i_2 \bullet} - y_{i_q \ldots i_3 \bullet\bullet}) \right]^2 + \ldots +$$

$$\ldots + \frac{I_1 \ldots I_{q-1} x_{q-3}}{1 - x_{q-2}} \sum_{i_q} \sum_{i_{q-1}} \left[b_{i_q i_{q-1}} - (1-x_{q-2}) \right.$$

$$\left. (y_{i_q i_{q-1} \ldots} - y_{i_q \bullet \ldots \bullet}) \right]^2 + \frac{I_1 \ldots I_{q-1} \; x_1 \ldots x_{q-2}}{1 - x_{q-1}} \sum_{i_q} \left[a_{i_q} - \right.$$

$$\left. - (1-x_{q-1}) (y_{i_q \bullet \ldots \bullet} - y_{\bullet \ldots \bullet}) \right]^2$$

(dots indicate summation over the respective indices) and

$$s + 1 = \frac{1}{2}(\nu_1 + 2\nu_2 + \ldots + 2\nu_q + 2).$$

Let

$$\hat{a}_{i_q}(x_{q-1}) = (1-x_{q-1}) (y_{i_q \bullet \ldots \bullet} - y_{\bullet \ldots \bullet})$$

$$(3.6) \qquad \hat{b}_{i_q i_{q-1}}(x_{q-2}) = (1-x_{q-2}) (y_{i_q i_{q-1} \bullet \ldots \bullet} - y_{i_q \bullet \ldots \bullet})$$

$$\vdots$$

$$\hat{d}_{i_q \ldots i_2}(x_1) = (1-x_1) (y_{i_q \ldots i_2 \bullet} - y_{i_q \ldots i_3 \bullet\bullet})$$

$$V = \nu_1 m_1 + x_1\nu_2 m_2 + \ldots + x_1\ldots x_{q-1}\nu_q m_q$$

$$a^*_{i_q} = [I_{q-1}\ldots I_1 V^{-1}(x_1\ldots x_{q-2})^{-1}(1-x_{q-1})^{-1}]^{\frac{1}{2}}[a_{i_q} - \hat{a}_{i_q}(x_{q-1})]$$

$$b^*_{i_q i_{q-1}} = [I_{q-2}\ldots I_1 V^{-1}(x_1\ldots x_{q-3})^{-1}(1-x_{q-2})^{-1}]^{\frac{1}{2}}[b_{i_q i_{q-1}} - \hat{b}_{i_q i_{q-1}}(x_{q-2})]$$

.
.
.

$$d^*_{i_q\ldots i_2} = [I_1 V^{-1}(1-x_1)^{-1}]^{\frac{1}{2}}[d_{i_q\ldots i_2} - \hat{d}_{i_q\ldots i_2}(x_1)].$$

Then integrating out σ_1^2 from (3.4) we obtain the density function of posterior distribution of $\underset{\sim}{a}^*, \ldots \underset{\sim}{d}^*$ given $x_1,\ldots,x_{q-1}$ and $\underset{\sim}{y}$ is

(3.7) $$f(\underset{\sim}{a}^*, \underset{\sim}{b}^*, \ldots, \underset{\sim}{d}^* | x_1, x_2, \ldots, x_{q-1}, \underset{\sim}{y}) =$$

$$= c\left[1 + \sum_{i_q} a^{*2}_{i_q} + \sum_{i_q}\sum_{i_{q-1}} b^{*2}_{i_q i_{q-1}} + \ldots + \sum_{i_q}\ldots\sum_{i_2} d^{*2}_{i_q\ldots i_2}\right]^{-s}$$

which is a multivariate t-distribution with $\sum_{m=1}^{q}\nu_m$ degrees of freedom.

Consequently

$$E(a_{i_q} | x_1\ldots x_{q-1}, \underset{\sim}{y}) = \hat{a}_{i_q}(x_{q-1})$$

$$E(b_{i_q i_{q-1}} | x_1 \ldots x_{q-1}, \underset{\sim}{y}) = \hat{b}_{i_q i_{q-1}}(x_q - 2)$$

$\cdot$

$\cdot$

$\cdot$

$$E(d_{i_q \ldots i_2} | x_1 \ldots x_{q-1}, \underset{\sim}{y}) = \hat{d}_{i_q \ldots i_2}(x_1)$$

$$\text{Var}(a_{i_q} | x_1 \ldots x_{q-1}, \underset{\sim}{y}) = \frac{(I_q - 1)\ (1 - x_{q-1}) V}{I_q \ldots I_1 x_1 \ldots x_{q-2} (\sum_{m=1}^{q} \nu_m - 2)}$$

$$\text{Var}(b_{i_q i_{q-1}} | x_1 \ldots x_{q-1}, \underset{\sim}{y}) = \frac{(I_{q-1} - 1)(1 - x_{q-2}) V}{I_{q-1} \ldots I_1 x_1 \ldots x_{q-3} (\sum_{m=1}^{q} \nu_m - 2)}$$

$\cdot$

$\cdot$

$\cdot$

$$\text{Var}\ (d_{i_q} \cdots {}_{i_2} | x_1 \ldots x_{q-1}, \underset{\sim}{y}) = \frac{(I_2 - 1)\ (1 - x_1) V}{I_2 I_1 (\sum_{m=1}^{q} \nu_m - 2)}$$

$$\text{Cov}(a_{i_q}, a_{i_q'} | x_1 \ldots x_{q-1}, \underset{\sim}{y}) = - \frac{1}{I_q - 1}\ \text{Var}(a_{i_q} | x_1 \ldots x_{q-1}, \underset{\sim}{y}),\ i_q \neq i_q'$$

$$\text{Cov}(b_{i_q i_{q-1}}, b_{i_q i'_{q-1}}, b_{i_q i_{q-1}} | x_1 \ldots x_{q-1}, \underset{\sim}{y}) = - \frac{1}{I_{q-1} - 1}\ \text{Var}(b_{i_q i_{q-1}} | x_1 \ldots$$

$$\ldots x_{q-1}, \underset{\sim}{y}), \qquad i_{q-1} \neq i'_{q-1}$$

$\cdot$

$\cdot$

$\cdot$

$$\text{Cov}(d_{i_q \ldots i_2}, d_{i_q \ldots i_2'} | x_1 \ldots x_{q-1}, \underset{\sim}{y}) = - \frac{1}{I_2 - 1}\ \text{Var}(d_{i_q \ldots i_2} / x_1 \ldots$$

$$\ldots x_{q-1} | \underset{\sim}{y}) \qquad i_2 \neq i_2'$$

The other covariances are equal zero.

Using well know results (see Raiffa and Schlaiffer (1961)) related to the multivariate t-distribution one can see that

$$E(a_{i_q}|\tilde{y}) = \underset{x_1\ldots x_{q-1}}{E}\left[E(a_{i_q}|\, x_1\ldots x_{q-1},\tilde{y})\right|\tilde{y}] = \underset{x_1\ldots x_{q-1}}{E}\left[\hat{a}_{i_q}(x_{q-1})|\,\tilde{y}\right]$$

$$(3.8) = (y_{i_y \bullet\ldots\bullet} - y_{\bullet\ldots\bullet})\ E\left[(1-x_{q-1})|\tilde{y}\right] = (1-\mu_{0\ldots 0,1})(y_{i_q \bullet\ldots\bullet}$$

$$- y_{\bullet\ldots\bullet}) = \hat{a}_{i_q}$$

where $\mu_{0\ldots o,1}$ is given by (2.5). ($\underset{x_1\ldots x_{q-1}}{E}[\ldots|.]$ denotes expectation with respect to $x_1\ldots x_{q-1}$).

In a similar fashion

$$E(b_{i_q i_{q-1}}|\tilde{y}) = (y_{i_q i_{q-1}\bullet\ldots\bullet} - y_{i_q \bullet\ldots\bullet})(1-\mu_{0\ldots 0,1,0}) = \hat{b}_{i_q i_{q-1}}$$

$$\vdots$$

$$E(d_{i_q\ldots i_2}|\,\tilde{y}) = (y_{i_q\ldots i_2 \bullet} - y_{i_q\ldots i_3 \bullet\bullet})\ (1-\mu_{1,0\ldots 0}) = \hat{d}_{i_q\ldots i_2}$$

Also

$$Var(a_{i_q}|\tilde{y}) = \underset{x_1\ldots x_{q-1}}{E}\left[Var(a_{i_q}|x_1\ldots x_{q-1},\tilde{y})\right] + \underset{x_1\ldots x_{q-1}}{E}\left[\hat{a}_{i_q}(x_{q-1}) -\right.$$

$$\left. - \hat{a}_{i_q}\right]^2 =$$

$$= \frac{I_{q-1}}{I_q \ldots I_1 \left(\sum_{m=1}^{q} \nu_m - 2 \right)} \nu_1 m_1 (\mu_{-1 \ldots -1,0} - \mu_{-1 \ldots -1,1}) + \nu_2 m_2 (\mu_{0,-1 \ldots -1,0} -$$

$$- \mu_{0,-1 \ldots -1,1}) + \nu_3 m_3 (\mu_{0,0,-1 \ldots -1,0} - \mu_{o,o,-1, \ldots, -1,1}) + \ldots +$$

$$+ \nu_q m_{q-1} (1 - \mu_{0 \ldots 0,1}) + \nu_q m_q (\mu_{0 \ldots 0,1} - \mu_{0 \ldots 0,2}) + (y_{i_q \bullet \ldots \bullet} - y_{\bullet \ldots \bullet})^2$$

$$(\mu_{0 \ldots 0,2} - \mu^2_{0 \ldots 0,1}).$$

The other variances and covariances we can write in an analogous way.

REFERENCES

BOX, G.E.P. and TIAO,G.C. (1965), Multivariate Problems from a Bayesian Point of view, Ann. Math. Statist. 36, pp.1468-1482.

CLEMENT, B. (1975), Estimation bayesianne dans le modéle classification double avec interaction et effets aléatoires, Revue Canadienne de Statistique, 3, pp. 203-221.

CLEMENT, B. (1977), Bayesian Estimation of Means in a Three Component Hierarchical Design with Random Effect Model, Statist. Hefte, 18, pp. 181-192.

GIRI, N.C. (1970), Bayesian Estimation of Means of a Two-Way Classification with Random Effect Model, Statist. Hefte 11, pp. 254-260.

RAIFFA, H. and SCHLAIFFER R. (1961), Applied Statistical Decision Theory, Harvard Graduate Business School.

TIAO, G.C. and BOX, G.E.P. (1967), Bayesian Analysis of a Three-Componenet Hierarchical Design Model, Biometrika, 54, pp. 109-125.

TIAO, G.C. and BOX, G.E.D. (1967), Bayesian Analysis of a Three-Component Hierarchical Design Model, Biometrika, 54, pp. 109-125.

TIAO, G.C. and BOX, G.E.P. (1968), Bayesian Estimation of Means for the Random Effect Model, J. Amer. Statist. Assoc. 63, pp. 174-181.

Dr. Helena Jelenkowska
Department of Applied Mathematics
Agricultural Academy of Lublin
Akademicka 13
PL-20-934 Lublin
Poland

Proc. of the 4th Pannonian Symp. on Math. Stat., Bad Tatzmannsdorf, Austria 1983
W. Grossmann, G. Pflug, I. Vincze, W. Wertz, eds

THE MOMENTS OF THE TESTSTATISTICS $D_{n,n}$ AND $D^+_{n,n}$ OF THE KOLMOGOROV-SMIRNOV TEST AND THEIR ASYMPTOTIC BEHAVIOR

WALTER KATZENBEISSER, WOLFGANG PANNY

Wien

0. INTRODUCTION

This paper deals with the moments of the teststatistics $D^+_{n,n}$ and $D_{n,n}$ of the one- and two sided two sample Kolmogorov-Smirnov test with equal sample sizes n under the null hypothesis. It is well known that there is a one-to-one correspondence between $D^+_{n,n}$, $D_{n,n}$ and the maximal one-, two sided deviation of a discrete path in the (x,y)-plane leading from (0,0) to (2n,0): in the k-th step of this path we move one unit up and to the right if the k-th smallest observation in the combined arrangement of all observations belongs to the first sample and one unit down and to the right if it belongs to the second sample. The distributions of the teststatistics under the null hypothesis is derived by considering all $\binom{2n}{n}$ paths as equally likely. The classical Kolmogorov-Smirnov teststatistics can be interpreted as suitably standardized deviations of such paths where the actual deviation is divided by n, the largest deviation that possibly can occur.

Throughout this paper we use the symbols $^+D_{2n}$ and D_{2n} to denote the corresponding "path statistics". There is an extremely large literature on the Kolmogorov-Smirnov test; we only want to quote /Dwass, 1967/, /Hajek, Sidak, 1967/ and /Durbin, 1973/.

Our main concern is the analysis of the asymptotic behavior of $E\{{}^{+}D_{2n}^{s}\}$ and $E\{D_{2n}^{s}\}$ where the connection to the moments fo the traditional statistics is $E\{D_{n,n}^{+s}\} = E\{{}^{+}D_{2n}^{s}\}/n^{s}$, $E\{D_{n,n}^{s}\} = E\{D_{2n}^{s}\}/n^{s}$. To derive the corresponding asymptotic formulae we rely on the so called Γ-function method conceived by De Bruijn. Our results may be regarded as supplements to the work of Gnedenko and Kemperman concerning the asymptotic behavior of the related distribution functions.

1. THE MOMENTS ${}^{+}D_{2n}^{s}$ OF THE ONE SIDED STATISTIC

It is well known that

$$P\{{}^{+}D_{2n} \geq h\} = \frac{\binom{2n}{n-h}}{\binom{2n}{n}}, \tag{1}$$

which is easily seen by simple reflection arguments. Using (1) the s-th moment can be written as

$$E\{{}^{+}D_{2n}^{s}\} = \sum_{m\geq 1} (m^{s}-(m-1)^{s}) \binom{2n}{n-m} / \binom{2n}{n} = \tag{2}$$

$$= \sum_{\lambda=0}^{s-1} (-1)^{s+\lambda+1} \binom{s}{\lambda} \sum_{m\geq 1} m^{\lambda} \binom{2n}{n-m} / \binom{2n}{n}. \tag{3}$$

The first representation can be derived by summation by parts. The second representation will be used in the sequel and is easily seen by the binomial theorem. For s=1,2,3 combinatorial identities are known /Riordan, 1968; p. 34/ which enable us to give "closed form" expressions for the corresponding moments:

$$E\{{}^{+}D_{2n}\} = \frac{1}{2}(-1+2^{2n}/\binom{2n}{n}),$$

$$E\{{}^{+}D_{2n}^2\} = \frac{1}{2}((2n+1) - 2^{2n}/\binom{2n}{n}),$$

$$E\{{}^{+}D_{2n}^3\} = \frac{1}{2}(-(3n+1) + (\frac{3}{2}n+1)2^{2n}/\binom{2n}{n}),$$

$$\mathrm{var}\{{}^{+}D_{2n}\} = \frac{1}{4}((4n+1)-2^{4n}/\binom{2n}{n}^2).$$

Let us start our asymptotic analysis with an investigation of the asymptotic behavior of $\binom{2n}{n-h}/\binom{2n}{n}$:

Lemma 1: For all $\varepsilon>0$, $m>0$

$$\binom{2n}{n-h}/\binom{2n}{n} = (1-\frac{1}{6n}(L^2-3L)+\frac{1}{48n^2}(\frac{2}{3}L^4 - \frac{36}{5}L^3+18L^2-8L))e^{-L} + \\ + \begin{cases} o(n^{-5/2}) & h<n^{1/2+\varepsilon} \\ o(n^{-m}) & h\geq n^{1/2+\varepsilon} \end{cases}$$

as $n\to\infty$, where $L= h^2/n$.

Proof: Using Stirling's formula we have

$$\binom{2n}{n-h}/\binom{2n}{n} = \frac{\Gamma^2(n+1)}{\Gamma(n-h+1)\ \Gamma(n+h+1)} =$$

$$= \exp\{-n(x^2+\frac{1}{2\cdot 3}x^4+ \frac{1}{3\cdot 5}x^6 +\frac{1}{4\cdot 7}x^8 +\dots) +$$

$$+\frac{1}{2}(x^2 +\frac{1}{2}x^4 + \frac{1}{3}x^6 +\dots) -$$

$$-\frac{1}{6n}(x^2+x^4 + \dots) + o(n^{-3})\},$$

where $x=h/n$ and the occuring power series converge for $|x|<1$. Confining now to $h<n^{1/2+\varepsilon}$, $\varepsilon>0$, we get

$$\exp\{-nx^2\}\exp\{-n(\tfrac{1}{6}x^4+\tfrac{1}{15}x^6)+(\tfrac{1}{2}x^2+\tfrac{1}{4}x^4)-\tfrac{1}{64}x^2\}\,\mathcal{O}(\exp\{n^{-3+\varepsilon}\}).$$

Expanding and arranging in terms of $L=h^2/n$ the assertion follows for $h<n^{1/2+\varepsilon}$. It only remains to consider the case $h\geq n^{1/2+\varepsilon}$. Applying again Stirling's formula to estimate the order of magnitude of $\binom{2n}{n-h}/\binom{2n}{n}$ one easily obtains

$$\binom{2n}{n-h} \Big/ \binom{2n}{n} = o(\exp\{-n^{2\varepsilon}\}).$$

The second error term of Lemma 1 follows by observing that the right hand side of the formula given in Lemma 1 is exponentially small, too. □

Lemma 1 could alternatively be obtained by a general result on lattice distributions due to Esseen /Esseen, 1945; p. 63/. We would like to mention further that by a more detailed analysis the error term could be improved to $o(n^{-3})$. The formulation of the second part of Lemma 1 may seem somewhat clumsy. However, it will prove convenient in the sequel.

Let us now introduce the functions

$$r_\lambda(n) = \sum_{m\geq 1} m^\lambda \exp(-m^2/n).$$

By means of the $r_\lambda(n)$ functions and an application of Lemma 1 the moments of ${}^{+}D_{2n}$ can be written as:

$$E\{{}^{+}D_{2n}^{s}\} = \sum_{\lambda=0}^{s-1} (-1)^{s+\lambda+1} \binom{s}{\lambda} \left[r_\lambda(n) + \frac{1}{6n}\left(-\frac{1}{n^2} r_{\lambda+4}(n) + \frac{3}{n} r_{\lambda+2}(n)\right) + \right.$$

$$\left. + \frac{1}{48n^2}\left(\frac{2}{3n^4} r_{\lambda+8}(n) - \frac{36}{5n^3} r_{\lambda+6}(n) + \frac{18}{n^2} r_{\lambda+4}(n) - \frac{8}{n} r_{\lambda+2}(n)\right) \right] \quad (4)$$

$$+ o(n^{(s-5)/2+\varepsilon}).$$

The error term results from observing that

$$\sum_{\lambda=0}^{s-1} (-1)^{s+\lambda+1} \binom{s}{\lambda} \sum_{1 \le k \le n^{1/2+\varepsilon}} k^\lambda \, o(n^{-5/2}) = o(n^{(s-5)/2+\varepsilon}) \; ;$$

on the other hand

$$\sum_{k \ge n^{1/2+\varepsilon}}^{n} k^\lambda \, o(n^{-m}) = o(n^{\lambda+1-m})$$

for all $m>0$. For $k>n$ the error is exponentially small and the given error term follows.

By (4) the analysis of the asymptotic behavior of the moments of ${}^{+}D_{2n}$ reduces now to the derivation of asymptotic equivalents for the $r_\lambda(n)$ functions.

Lemma 2: The asymptotic behavior of $r_\lambda(n)$, $\lambda=0,1,2,\ldots$, can be described by

$$r_\lambda(n) = \frac{1}{2}\Gamma\left(\frac{\lambda+1}{2}\right) n^{(\lambda+1)/2} + \sum_{k=0}^{m-1} \frac{(-1)^k}{k!} \zeta(-2k-\lambda) n^{-k} + o(n^{-m})$$

for all $m>0$ as $n \to \infty$.

Proof: The above expansion could be obtained by Euler's summation formula. We prefer to use the so called Γ-function method (cf. /Knuth, 1973; p. 130f/). Starting from the well known formula (cf. e.g. /Whittaker, Watson, 1927; p. 280/)

$$e^{-x} = \frac{1}{2\pi i} \int_{c-i\infty}^{c+i\infty} \Gamma(z) x^{-z} dz, \; c>0, \; x>0, \tag{5}$$

and regarding that

$$\sum_{k\geq 1} \frac{k^{\lambda}}{k^{z}} = \zeta(z-\lambda),$$

we find

$$r_{\lambda}(n) = \sum_{k\geq 1} \frac{1}{2\pi i} \int_{c-i\infty}^{c+i\infty} \frac{n^{z}}{k^{2z-\lambda}} \Gamma(z) dz = \frac{1}{2\pi i} \int_{c-i\infty}^{c+i\infty} n^{z} \Gamma(z) \zeta(2z-\lambda) dz,$$

where now $c>(\lambda+1)/2$. This integral representation can be taken as a starting point to derive asymptotic equivalents for the $r_{\lambda}(n)$-functions as $n\to\infty$, essentially by applying the residue theorem. Considering the asymptotic behavior of the involved Γ- and ζ- functions (cf. /Abramowitz, Stegun, 1972; p. 257/, /Whittaker, Watson, 1927; p. 276/, resp.) we see that the line of integration may be shifted to the left as far as we please if we only take the residues into account, yielding an error of $o(n^{-m})$ for all $m>0$ as $n\to\infty$. At $z=(\lambda+1)/2$ we have a simple pole with residue $\frac{1}{2}\Gamma(\frac{\lambda+1}{2}) n^{(\lambda+1)/2}$, due to Riemann's ζ-function. At $z=-k$, $k\in\mathbb{N}_0$, there are possibly simple poles with residues $(-1)^k \zeta(-2k-\lambda) n^{-k}/k!$, caused by the poles of the Γ-function.

Summing up gives an asymptotic series for $r_{\lambda}(n)$. □

so we obtain:

$$r_o(n) = \frac{1}{2}(\sqrt{\pi}n^{1/2} - 1) + o(n^{-m}),$$

$$r_\lambda(n) = \begin{cases} \frac{1}{2}\sqrt{\pi}\frac{1\cdot 3\cdot 5\cdot \;...\;\cdot(\lambda-1)}{2^{\lambda/2}} n^{(\lambda+1)/2} + o(n^{-m}), & \lambda=2,4,.. \\ \\ \frac{1}{2}(\frac{\lambda-1}{2})!n^{(\lambda+1)/2} + \sum_{k=0}^{m-1}\frac{(-1)^k(-B_{2k+\lambda+1})}{k!n^k(2k+\lambda+1)} + o(n^{-m}), & \lambda=1,3,.. \end{cases} \tag{6}$$

as $n\to\infty$, where B_k denotes the k-th Bernoulli number.

Returning to our original problem about the moments of $^+D_{2n}$ we finally can state the following

<u>Theorem 1:</u> For all $\varepsilon>0$ we have as $n\to\infty$:

$$E\{^+D_{2n}\} = \frac{1}{2}\sqrt{\pi}\,n^{1/2} - \frac{1}{2} + \frac{1}{16}\sqrt{\pi}\,n^{-1/2} + \frac{1}{256}\sqrt{\pi}\,n^{-3/2} + o(n^{-2+\varepsilon}),$$

$$E\{^+D_{2n}^2\} = n - \frac{1}{2}\sqrt{\pi}n^{1/2} + \frac{1}{2} - \frac{1}{16}\sqrt{\pi}\,n^{-1/2} + o(n^{-3/2+\varepsilon}),$$

$$E\{^+D_{2n}^3\} = \frac{3}{4}\sqrt{\pi}\,n^{3/2} - \frac{3}{2}n + \frac{19}{32}\sqrt{\pi}\,n^{1/2} - \frac{1}{2} - \frac{469}{512}\sqrt{\pi}\,n^{-1/2} + o(n^{-1+\varepsilon}),$$

$$E\{^+D_{2n}^4\} = 2n^2 + 2n - \frac{3}{2}\sqrt{\pi}n^{3/2} - \frac{19}{16}\sqrt{\pi}\,n^{1/2} + \frac{1}{2} + o(n^{-1/2+\varepsilon}),$$

$$E\{^+D_{2n}^s\} = \frac{1}{2}\sum_{\lambda=0}^{4}(-1)^\lambda\binom{s}{\lambda+1}\Gamma(\frac{s-\lambda}{2})n^{(s-\lambda)/2} -$$

$$- \frac{1}{24}\sum_{\lambda=0}^{2}(-1)^\lambda\binom{s}{\lambda+1}(s-4-\lambda)\,\Gamma(\frac{s+2-\lambda}{2})\,n^{(s-2-\lambda)/2} +$$

$$+ \frac{s}{5760} (5s^3 - 48s^2 + 112s - 24)\ \Gamma\left(\frac{s+2}{2}\right) n^{(s-4)/2} +$$

$$+ \ o(n^{(s-5)/2+\varepsilon}), \qquad \text{for } s > 5.$$

$$\text{var}\ \{{}^{+}D_{2n}\} = (1 - \tfrac{1}{4}\pi)n + (\tfrac{1}{4} - \tfrac{\pi}{16}) - \tfrac{\pi}{128}n^{-1} + o(n^{-3/2+\varepsilon}). \quad \square$$

In the following table computational results are summarized, where asymptotic (as.) and exact (ex.) values are compared. They were computed using floating point arithmetic with a seven byte mantissa.

Table 1: Expectation and variance of ${}^{+}D_{2n}$

	$E\{{}^{+}D_{2n}\}$		$\text{var}\{{}^{+}D_{2n}\}$	
n	ex.	as.	ex.	as.
2	0.833333	0.834094	0.47222	0.47058
4	1.328571	1.328709	0.90633	0.90592
6	1.716450	1.716500	1.33735	1.33717
8	2.046076	2.046100	1.76750	1.76740
10	2.337732	2.337746	2.19728	2.19721
15	2.961059	2.961064	3.27107	3.27104
20	3.488173	3.488175	4.34448	4.34446
25	3.953344	3.953346	5.41772	5.41771
50	5.782256	5.782257	10.78325	10.78325
75	7.187752	7.187753	16.14846	16.14846
100	8.373354	8.373354	21.51359	21.51359

2. THE MOMENTS D_{2n}^s OF THE TWO SIDED STATISTIC

In this section we proceed in a similar way as in section 1. So we start from the well known fact

$$P\{D_{2n} \geq h\} = 2 \sum_{\lambda \geq 1} (-1)^{\lambda+1} \frac{\binom{2n}{n-\lambda h}}{\binom{2n}{n}}, \tag{7}$$

which can be derived by the method of repeated reflections /Feller, 1968; p. 96/. Hence

$$E\{D_{2n}^s\} = 2 \sum_{h \geq 1} (h^s - (h-1)^s) \sum_{\lambda \geq 1} (-1)^{\lambda+1} \binom{2n}{n-\lambda h} / \binom{2n}{n}, \tag{8}$$

which is easily obtained using summation by parts again. If we introduce the arithmetical functions

$$\tau_s(m) = \sum_{\substack{m=\lambda h \\ \lambda=1,3,..}} h^s \qquad \text{and} \qquad \sigma_s(m) = \sum_{\substack{m=\lambda h \\ \lambda=2,4,..}} h^s ,$$

(8) can be written as:

$$E\{D_{2n}^s\} = 2 \sum_{\lambda=0}^{s-1} (-1)^{s+1-\lambda} \binom{s}{\lambda} \sum_{m \geq 1} (\tau_\lambda(m) - \sigma_\lambda(m)) \binom{2n}{n-m} / \binom{2n}{n}. \tag{9}$$

Let us now introduce the functions

$$q_{b,s}(n) = \sum_{m \geq 1} (\tau_s(m) - \sigma_s(m)) m^b \exp(-m^2/n),$$

which constitute the analogues to the r-functions of the preceding section. Using Lemma 1 the $q_{b,s}$ functions enable us to express (9) as:

$$E\{D_{2n}^{s}\} = \sum_{\lambda=0}^{s-1}(-1)^{s+\lambda+1}\binom{s}{\lambda}\Big[2q_{0,\lambda}(n) - \frac{1}{3n}\Big(\frac{1}{n^2}q_{4,\lambda}(n) - \frac{3}{n}q_{2\lambda}(n)\Big) + \tag{10}$$

$$+\frac{1}{24n^2}\Big(\frac{2}{3n^4}q_{8,\lambda}(n) - \frac{36}{5n^3}q_{6,\lambda}(n) + \frac{1}{18n^2}q_{4,\lambda}(n) - \frac{8}{n}q_{2,\lambda}(n)\Big)\Big] +$$

$$+ o(n^{(s-5)/2+\varepsilon}).$$

The o-term in (10) is obtained by dissecting according to $1\le k<n^{1/2+\varepsilon}$ and $k\ge n^{1/2+\varepsilon}$, respectively. For the first part we use the well known result $\sum_{1\le k\le n} d(k) = n \,\ell n\, n + o(n)$, $n\to\infty$, due to Dirichlet (cf. e.g. /Chandrasekharan, 1966/). Thus we obtain

$$\sum_{\lambda=0}^{s-1}(-1)^{s+\lambda+1}\binom{s}{\lambda}\sum_{1\le m\le n^{1/2+\varepsilon}}\tau_\lambda(m) \le \sum_{\lambda=0}^{s-1}(-1)^{s+\lambda+1}\binom{s}{\lambda}\sum_{1\le m\le n^{1/2+\varepsilon}}\sum_{h|m} h^{\lambda} \le$$

$$\le \sum_{\lambda=0}^{s-1}(-1)^{s+\lambda+1}\binom{s}{\lambda} n^{\lambda(1/2+\varepsilon)}\sum_{1\le m\le n^{1/2+\varepsilon}} d(m) = o(n^{s/2+\varepsilon}\ell n\, n).$$

For the contribution of the $\sigma_\lambda(m)$ functions a similar argument applies and the first part is $o(n^{(s-5)/2+\varepsilon})$, where ln n is omitted regarding ε. The contribution of the second part may be neglected being only $o(n^{-m})$ for all $m>0$ as $n\to\infty$.

Now our problem boils down to the investigation of the asymptotic behavior of the $q_{b,\lambda}(n)$ functions. To accomplish this task we again resort to the Γ-function method.

Lemma 3: For b=0,2,4, ... the asymptotic behavior of $q_{b,s}(n)$ can be described by

$$q_{b,0}(n) = \frac{1}{2}(\ln 2)\Gamma\left(\frac{b+1}{2}\right)n^{(b+1)/2} + o(n^{-m}) - \begin{cases} 1/4 & b=0 \\ 0 & \text{otherwise} \end{cases} \tag{11a}$$

for all m>0 as $n\to\infty$. If s>0 we have

$$q_{b,s}(n) = \frac{1}{2}(1-2^{-s})\zeta(1+s)\Gamma\left(\frac{b+s+1}{2}\right)n^{(b+s+1)/2} + o(n^{-m}) - \begin{cases} \frac{B_{s+1}}{2(s+1)} & b=0 \wedge s\equiv 1(2) \\ 0 & \text{otherwise} \end{cases} \tag{11b}$$

for all m>0 as $n\to\infty$, where B_k denotes the k-th Bernoulli number and the values of the ζ- and Γ- functions are well known for the special arguments occuring in (11) (cf.e.g. /Abramowitz, Stegun, 1972/).

Proof: By (5)

$$q_{b,s}(n) = \sum_{m\geq 1}(\tau_s(m)-\sigma_s(m))m^b\exp(-m^2/n) =$$

$$= \sum_{m\geq 1}(\tau_s(m)-\sigma_s(m))\frac{1}{2\pi i}\int_{c-i\infty}^{c+i\infty} m^b\Gamma(z)\,(m^2/n)^{-z}dz,\ c>\frac{b+s+1}{2}.$$

The sum may be placed inside the integral, since convergence is absolutely well behaved, and thus the last expression equals

$$\frac{1}{2\pi i}\int_{c-i\infty}^{c+i\infty}(1-2^{-2z+b+1})\zeta(2z-b)\zeta(2z-b-s)\Gamma(z)n^{z}dz,$$

which is seen by observing that the Dirichlet series of $\sigma_s(m)$, $\tau_s(m)$ are

$$\sum_{m\geq 1}\frac{\sigma_s(m)}{m^z} = 2^{-z}\zeta(z)\zeta(z-s), \quad \sum_{m\geq 1}\frac{\tau_s(m)}{m^z} = (1-2^{-z})\zeta(z)\zeta(z-s).$$

By the cited method it can be shown that the line of integration may be shifted to the left as far as we please if we only take the residues into account, yielding an error of $o(n^{-m})$ for all $m>0$ as $n\to\infty$.

Possibly there are some simple poles at $z=-k$, $k=0,1,2,\ldots,$ with residues

$$\frac{(-1)^k}{k!}(1-2^{2k+b+1})\zeta(-2k-b)\zeta(-2k-b-s)n^{-k},$$

due to the Γ-function. If $s=0$ we have a simple pole at $z=(b+1)/2$, owing to $\zeta^2(2z-b)$. Since

$$(1-2^{-2z+b+1})\zeta^2(2z-b)\Gamma(z)n^z =$$

$$= (2w\,\ln 2+o(w^2))\left(\tfrac{1}{4}w^{-2}+\tfrac{\gamma}{w}+o(1)\right)\left(\Gamma\left(\tfrac{b+1}{2}\right)+o(w)\right)\left(n^{(b+1)/2}+o(w)\right),$$

where $w=z-(b+1)/2$, the residue at this pole is

$$\tfrac{1}{2}\Gamma\left(\tfrac{b+1}{2}\right)(\ln 2)\; n^{(b+1)/2}.$$

If $s>0$ there is a simple pole at $z=(1+b+s)/2$ with residue

$$\frac{1}{2}(1-2^{-s})\zeta(1+s)\Gamma(\frac{1+b+s}{2})n^{(1+b+s)/2},$$

caused by $\zeta(2z-b-s)$. Summing up and specializing on $b\equiv 0(2)$ yields the asymptotic equivalents given above. □

Let us summarize the result of this section in the following

<u>Theorem 2:</u> For all $\varepsilon>0$ we have as $n\to\infty$:

$$E\{D_{2n}\} = \sqrt{\pi}(\ell n2)n^{1/2} - \frac{1}{2} + \frac{1}{8}\sqrt{\pi}(\ell n2)n^{-1/2} + \frac{1}{128}\sqrt{\pi}(\ell n2)n^{-3/2} + o(n^{-2+\varepsilon}),$$

$$E\{D_{2n}^2\} = \frac{\pi^2}{6}n - \sqrt{\pi}(\ell n2)n^{1/2} + \frac{1}{3} + \frac{\pi^2}{36} - \frac{1}{8}\sqrt{\pi}(\ell n2)n^{-1/2} + \frac{\pi^2}{360}n^{-1} + o(n^{-3/2+\varepsilon}),$$

$$E\{D_{2n}^3\} = \frac{9}{8}\sqrt{\pi}\zeta(3)n^{3/2} - \frac{\pi^2}{4}n + \sqrt{\pi}(\frac{9}{64}\zeta(3)+\ell n2)n^{1/2} - \frac{1}{4} - \frac{\pi^2}{24} -$$

$$- \frac{\sqrt{\pi}}{1024}(1503\zeta(3)-128\ell n2)n^{-1/2} + o(n^{-1+\varepsilon}),$$

$$E\{D_{2n}^4\} = \frac{7\pi^4}{180}n^2 - \frac{9}{4}\sqrt{\pi}\zeta(3)n^{3/2} + \frac{\pi^2}{3}n - \sqrt{\pi}(\ell n2+\frac{9}{32}\zeta(3))n^{1/2} +$$

$$+ \frac{1}{10800}(-7\pi^4 + 600\pi^2 + 2160) + o(n^{-1/2+\varepsilon}),$$

$$E\{D_{2n}^s\} = \sum_{\lambda=}^{4}(-1)^{\lambda}\binom{s}{\lambda+1}(1-2^{\lambda+1-s})\zeta(s-\lambda)\Gamma(\frac{s-\lambda}{2})n^{(s-\lambda)/2} +$$

$$+ \sum_{\lambda=0}^{2}(-1)^{\lambda}\binom{s}{\lambda+1}\frac{s-4-\lambda}{12}(2^{1+\lambda-s}-1)\zeta(s-\lambda)\Gamma(\frac{s+2-\lambda}{2})n^{(s-2-\lambda)/2} +$$

$$+ s\frac{5s^3-48s^2+112s-24}{2880}(1-2^{1-s})\zeta(s)\Gamma(\frac{s+2}{2})n^{(s-4)/2}+O(n^{(s-5)/2+\varepsilon}),\ s\geq 5.$$

$$\mathrm{Var}\{D_{2n}\} = \pi(\frac{\pi}{6}-(\ell n2)^2)n + \frac{1}{36}(3-9\pi(\ell n2)^2+\pi^2) +$$

$$+ \frac{\pi}{1440}(4\pi-45(\ell n2)^2)n^{-1} + O(n^{-3/2+\varepsilon}). \square$$

As in the preceding section we give some computational results shown in Table 2:

Table 2: Expectation and variance of D_{2n}

	$E\{D_{2n}\}$		$var\{D_{2n}\}$	
n.	ex.	as.	ex.	as.
2	1.333333	1.349447	0.22222	0.24136
4	2.028571	2.035128	0.54204	0.51739
6	2.575757	2.572721	0.78539	0.79013
8	3.031235	3.029644	1.05435	1.06204
10	3.433805	3.433951	1.33215	1.33363
15	4.297838	4.298053	2.01367	2.01202
20	5.028877	5.028785	2.68961	2.69008
25	5.673670	5.673648	3.36773	3.36801
50	8.209059	8.209056	6.75704	6.75707
75	10.157488	10.157487	10.14586	10.14586
100	11.801081	11.801080	13.53459	13.53459

3. CONCLUDING REMARKS

Following the method applied in Lemma 1 we could improve the accuracy of this approximation by using more terms of the underlying expansion. So the asymptotic expansions of Theorems 1 and 2 could be extended to achieve an error term as small as we please. By the same arguments it can be shown that the ε in the error terms of Theorems 1 and 2 are actually unnecessary and could be omitted.

The computational results suggest that the asymptotic approximations derived in this paper can be used even for small values of n.

It seems worth to mention that (1) and (7) can alternatively be obtained by purely analytical arguments, following the method used in /de Bruijn, Knuth, Rice, 1972/, /Prodinger, 1980/, /Panny, 1981/. In Prodinger's paper also the asymptotic equivalent for the first moment of D_{2n} was derived (with an error term $O(n^{-1/2})$).

REFERENCES

M.ABRAMOWITZ and I. STEGUN, ed., Handbook of Mathematical Functions, 10th printing, Applied Mathematics Series 55 (National Bureau of Standards, 1972).

N.G.DE BRUIJN, D.E.KNUTH and S.O. RICE, The average height of planted plane trees, in:R.C. Read, ed., Graph Theory and Computing (Academic Press, New York, 1972) 15-22.

K. CHANDRASEKHARAN, Einführung in die Analytische Zahlentheorie, Lecture Notes in Mathematics 29 (Springer, New York, 1966).

J. DURBIN, Distribution Theory for Tests Based on the Sample Distribution Function, Regional Conference Series in Applied Mathematics 9 (Siam, 1973).

M.DWASS, Simple random walk and rank order statistics, Ann. Math. Statist. 38. (1967) 1042-1053.

C.G.ESSEEN, Fourier analysis of distribution functions: A mathematical study of the Laplace-Gaussian law, Acta Math. 77 (1945) 3-125.

W.FELLER, An Introduction to Probability Theory and its Applications, Vol. I. 3rd ed. (Wiley, New York, 1968).

B.V.GNEDENKO, Some results on the maximal deviation between two empirical distributions, Dokl. Akad. Nauk SSSR, 82 (1952) 661-663.

J.HÁJEK and Z.SIDÁK, Theory of Rank Tests (Academic Press, New York, 1967).

J.H.B.KEMPERMAN, Asymptotic expansions for the Smirnov test and for the range of cumulative sums, Ann. Math. Statist. 30. (1959) 448-462.

D.E.KNUTH, The Art of Computer Programming, Vol. III. (Addison-Wesley, Reading/Mass., 1972).

W.PANNY, The expected depth of a pushdown store with three operations and symmetrical probabilities, J.Comb. Inf.Syst.Sci. 7 (1981) 38-47.

H.PRODINGER, The average height of a stack where three operations are allowed and some related problems, J.Comb.Inf.Syst.Sci. 5 (1980) 287-304.

J. RIORDAN, Combinatorial Identities (Wiley, New York, 1968).

E.T.WHITTAKER and G.N. WATSON, A Course of Modern Analysis, 4th ed. (Cambridge Universitiy Press, Cambridge, 1927).

W. Katzenbeisser
W. Panny

Institut für Statistik
Abteilung für mathematische Methoden
der Statistik

University of Economics, Vienna

Augasse 2-6.
A-1090 Vienna
Austria

Proc. of the 4th Pannonian Symp. on Math. Stat., Bad Tatzmannsdorf, Austria 1983
W. Grossmann, G. Pflug, I. Vincze, W. Wertz, eds

ESTIMATION ASSOCIATED WITH LINEAR AND QUADRATID DISCRIMINANT FUNCTIONS

MIROSLAW KRZYŚKO

Poznań

1. Introduction. Let us assume that in the population π_i, the observed random vector $\underline{X}_i$ has a p-dimensional normal distribution with the density function $f(\underline{x};\ \underline{\mu}_i,\ \Sigma_i)$, $\Sigma_i > 0$, $i = 1,2$. We have an observation $\underline{x}$, which we wish to assign to one of the two populations π_1 or π_2. On many approaches to the problem of discrimination thus formulated, we shall here select a decision-theoretic approach (see, e.g. Anderson, 1958, Chap. 6).

The quadratic discriminant function obtained within this approach is a function of the parameters $\underline{\mu}_i$ and $\underline{\Sigma}_i$ of the following form:

$$u(\underline{x}) = \ln(f(\underline{x};\ \underline{\mu}_1,\ \underline{\Sigma}_1)/f(\underline{x};\ \underline{\mu}_2,\ \underline{\Sigma}_2)) = \frac{1}{2}[\Delta_2^2(\underline{x}) - \Delta_1^2(\underline{x}) + \ln(|\underline{\Sigma}_2|/|\underline{\Sigma}_1|)] \tag{1}$$

where

$$\Delta_i^2(\underline{x}) = (\underline{x} - \underline{\mu}_i)'\underline{\Sigma}_i^{-1}(\underline{x} - \underline{\mu}_i), \quad i = 1,2.$$

The classification rule has the following form:
Assign $\underline{x}$ to π_1 if and only if $u(\underline{x}) > \ln(q_2/q_1)$, otherwise to π_2, where q_i is the prior probability that $\underline{x}$ comes from π_i, $i=1,2$.

In order to be able to use the quadratic discriminant function it is necessary to know the parameters $\underline{\mu}_i$ and $\underline{\Sigma}_i$, $i = 1,2$. If these parameters are not known, the following four estimators are used at present: the usual biased and unbiased estimators and two Bayesian estimators of the quadratic discriminant function. Some properties of these estimators are investigated here. Simulation is used in order to compare these estimators in the case of small samples.

2. The frequentistic approach to the estimation. Let $\underline{\bar{x}}_i$ and $\underline{S}_i$ denote the usual estimators of the parameters $\underline{\mu}_i$ and $\underline{\Sigma}_i$ obtained from two independent random samples of sizes N_1 and N_2 respectively, $N_1, N_2 \geq p+1$. If, in the quadratic discriminant function of the form (1), we replace the unknown parameters by their conventional estimators, we obtain the estimator of the quadratic discriminant function of the form:

$$\hat{u}_1(\underline{x}) = \frac{1}{2}[D_2^2(\underline{x}) - D_1^2(\underline{x}) + \ln(|\underline{S}_2|/|\underline{S}_1|)], \tag{2}$$

where

$$D_i^2(\underline{x}) = (\underline{x} - \underline{\bar{x}}_i)'\underline{S}_i^{-1}(\underline{x} - \underline{\bar{x}}_i), \quad i = 1,2.$$

We shall now consider the consistency and the unbiasedness of the estimator $\hat{u}_1(\underline{x})$ of the quadratic discriminant function $u(\underline{x})$.

We have

$$\operatorname*{plim}_{N_1,N_2\to\infty} \hat{u}_1(\underline{x}) = u(\underline{x}),$$

where plim denotes convergence in probability. Consequently, $\hat{u}_1(x)$ is a consistent estimator of the quadratic discriminant function $u(\underline{x})$. In the Appendix it is shown that the expected value of the estimator $\hat{u}_1(\underline{x})$ is:

$$E[\hat{u}_1(\underline{x})] = \frac{1}{2}\left[\frac{N_2-1}{N_2-p-2}\,\Delta_2^2(\underline{x}) - \frac{N_1-1}{N_1-p-2}\,\Delta_1^2(\underline{x}) + \right. \tag{3}$$

$$\left. + \ln(|\underline{\Sigma}_2|/|\underline{\Sigma}_1|) + h_1(p,N_1,N_2)+h_2(p,N_1,N_2)\right],$$

where

$$h_1(p,N_1,N_2) = \sum_{j=1}^{p}[\Psi(\tfrac{1}{2}(N_2-j)) - \Psi(\tfrac{1}{2}(N_1-j))] + p\ln\frac{N_1-1}{N_2-1} \tag{4}$$

$$h_2(p,N_1,N_2) = p\left(\frac{N_2-1}{N_2(N_2-p-2)} - \frac{N_1-1}{N_1(N_1-p-2)}\right), \tag{5}$$

$\Psi(x) = \dfrac{d\ln\Gamma(x)}{dx}$ is the psi (diagamma) function (see Abramowitz and Stegun, 1965, p. 258).

We see that

$$E[\hat{u}_1(\underline{x})] \neq u(\underline{x}).$$

We have

$$\lim_{N_1,N_2\to\infty} h_2(p,N_1,N_2) = 0.$$

Furthermore, since for large x the function $\Psi(x)$ is asymptotically equal to $\ln x + O(1)$, it is clear that

$$\lim_{N_1,N_2\to\infty} h_1(p,N_1,N_2) = 0$$

and

$$\lim_{N_1,N_2\to\infty} E[\hat{u}_1(\underline{x})] = u(\underline{x}).$$

Hence, the estimator $\hat{u}_1(\underline{x})$ is a biased estimator of the function $u(\underline{x})$, but asymptotically unbiased.

The unbiased estimator of the quadratic discriminant function has the following form:

$$\hat{u}_2(\underline{x}) = \frac{1}{2}\left[\frac{N_2-p-2}{N_2-1} D_2^2(\underline{x}) - \frac{N_1-p-2}{N_1-1} D_1^2(\underline{x}) + \right. \tag{6}$$

$$\left. + \ln(|\underline{S}_2|/|\underline{S}_1|) - h_1(p,N_1,N_2) - h_2(p,N_1,N_2)\right].$$

In addition to being unbiased the above estimator has minimum variance being a function of the complete sufficient statistics for the unknown parameters.

3. The Bayesian approach to the estimation. If the quadratic loss function is used, then the Bayes estimator of the quadratic discriminant function $u(\underline{x})$ is the expected value of that function with respect to the joint posterior distribution of the parameters which occur in it. We shall denote the estimator by $\hat{u}_3(\underline{x})$. Assuming that the density function of the joint prior distribution of the parameters $(\underline{\mu}_i, \underline{\Sigma}_i^{-1})$ is a Jeffreys (1961) function of the form

$$g(\underline{\mu}_i, \underline{\Sigma}_i^{-1}) \propto |\underline{\Sigma}_i|^{\frac{p+1}{2}}, \tag{7}$$

we obtain (see Appendix):

$$\hat{u}_3(\underline{x}) = \hat{u}_1(\underline{x}) - \frac{1}{2} h_1(p,N_1,N_2). \tag{8}$$

It is clear from (4) that $h_1(p,N,N) = 0$. Hence, if the sample sizes are equal, the Bayes estimator is equal to the usual biased estimator of the quadratic discriminant function.

We shall now consider the Bayes estimator of the density function $f(\underline{x};\ \underline{\mu}_i,\ \underline{\Sigma}_i)$, i = 1,2. If the quadratic loss function is used, then the Bayes estimator of the density function $f(\underline{x};\ \underline{\mu}_i,\ \underline{\Sigma}_i)$ is the expected value of that function with respect to the posterior distribution of the parameters which occur in it. We shall denote this estimator by $\hat{f}(\underline{x};\underline{\bar{x}}_i,\underline{S}_i)$, i = 1,2. We have

$$\hat{f}(\underline{x};\underline{\bar{x}}_i,\underline{S}_i) = \int\int f(\underline{x};\underline{\mu}_i,\underline{\Sigma}_i)t(\underline{\mu}_i,\underline{\Sigma}_i|\underline{\bar{x}}_i,\underline{S}_i)d\underline{\mu}_i d\underline{\Sigma}_i \ ,$$

where $t(\underline{\mu}_i,\underline{\Sigma}_i|\underline{\bar{x}}_i,\underline{S}_i)$ is the density function of the posterior distribution of the parameters $(\underline{\mu}_i,\underline{\Sigma}_i)$, i = 1,2. Assuming that the density function of the joint prior distribution of the parameters $(\underline{\mu}_i,\underline{\Sigma}_i)$ is a function of the form (7) we obtain (Geisser, 1966):

$$f(\underline{x};\underline{\bar{x}}_i,\underline{S}_i) = c_i \left[1 + \frac{N_i}{N_i^2-1} D_i^2(\underline{x})\right]^{-\frac{N_i}{2}} , \qquad (9)$$

where

$$c_i = \left(\frac{N_i}{\pi(N_i+1)}\right)^{\frac{p}{2}} \frac{\Gamma\left(\frac{N_i}{2}\right)}{\Gamma\left(\frac{N_i-p}{2}\right)|(N_i-1)\underline{S}_i|^{\frac{1}{2}}} , \ i = 1,2.$$

The function (9) is the density function of the p-dimensional t-distribution (Cornish, 1954).

If we use the estimator (9), then the Bayes estimator of the quadratic discriminant function takes the following form:

$$\hat{u}_4(\underline{x}) = \frac{N_2}{2} \ln [1 + N_2(N_2^2 - 1)^{-1}D_2^2(\underline{x})] - $$
$$- \frac{N_1}{2} \ln [1 + N_1(N_1^2 - 1)^{-1}D_1^2(\underline{x})] + \ln c_1c_2^{-1}. \tag{10}$$

Because of

$$\operatorname*{plim}_{N_i \to \infty} \{N_i \ln[1 + N_i(N_i^2 - 1)^{-1}D_i^2(\underline{x})]\} = \Delta_i^2(\underline{x}), \quad i=1,2$$

and

$$\operatorname*{plim}_{N_1,N_2 \to \infty} [\ln(c_1/c_2)^2] = \ln(|\underline{\Sigma}_2|/|\underline{\Sigma}_1|),$$

it follows that

$$\operatorname*{plim}_{N_1,N_2 \to \infty} \hat{u}_4(\underline{x}) = u(\underline{x})$$

holds.

Therefore, $\hat{u}_4(\underline{x})$ is also a consistent estimator of the quadratic discriminant function $u(\underline{x})$.

4. The estimators of the linear discriminant function

Let us assume that $\underline{\Sigma}_1 = \underline{\Sigma}_2 = \underline{\Sigma}$. Then the linear discriminant function is a function of the parameters $\underline{\mu}_i$ and $\underline{\Sigma}$ of the following form:

$$e(\underline{x}) = \frac{1}{2}[\delta_2^2(\underline{x}) - \delta_1^2(\underline{x})],$$

where

$$\delta_i^2(\underline{x}) = (\underline{x} - \underline{\mu}_i)'\underline{\Sigma}^{-1}(\underline{x} - \underline{\mu}_i), \quad i=1,2.$$

The estimators of the linear discriminant function corresponding to the estimators of the quadratic discriminant function are:

$$\hat{e}_1(\underline{x}) = \frac{1}{2}[d_2^2(\underline{x}) - d_1^2(\underline{x})], \tag{12}$$

where

$$d_i^2(\underline{x}) = (\underline{x} - \overline{\underline{x}}_i)' \ \underline{S}^{-1}(\underline{x} - \overline{\underline{x}}_i)$$

and $\underline{S}$ denotes the usual estimator of the matrix $\underline{\Sigma}$ obtained from two samples of sizes N_1 and N_2,

$$\hat{e}_2(\underline{x}) = \frac{1}{2}[k_1(p,N_1,N_2)(d_2^2(\underline{x}) - d_1^2(\underline{x})) + + k_1^{-1}(p,N_1,N_2)k_2(p,N_1,N_2)], \tag{13}$$

where

$$k_1(p,N_1,N_2) = \frac{N_1+N_2-p-3}{N_1+N_2-2}, \ k_2(p,N_1,N_2) = p\left(\frac{1}{N_1} - \frac{1}{N_2}\right),$$

$$N_1 + N_2 - p - 3 > 0,$$

$$\hat{e}_3(\underline{x}) = \hat{e}_1(\underline{x}) - \frac{1}{2} k_2(p,N_1,N_2), \tag{14}$$

$$\begin{aligned}\hat{e}_4(\underline{x}) = \frac{1}{2}\{&(N_1+N_2-3)\ell n[1+N_2(N_2+1)^{-1}(N_1+N_2-2)^{-1}d_2^2(\underline{x})] - \\ &- (N_1+N_2-3)\ell n[1+N_1(N_1+1)^{-1}(N_1+N_2-2)^{-1}d_1^2(\underline{x})] + \\ &+ p\ \ell n[N_1(N_2+1)/N_2(N_1+1)]\}.\end{aligned} \tag{15}$$

5. Simulation results of the comparison between the estimators

For a simulation study, we consider a situation with two populations where the vector distributions are p-dimensional multinormal $N_p(\underline{\mu}_1, \Sigma_1)$ and $N_p(\underline{\mu}_2, \underline{\Sigma}_2)$. Since we can find a non--singular linear transformation $\underline{L}$ such that $L\ \underline{\Sigma}_1\ \underline{L}' = \underline{I}_p$, $L\ \underline{\Sigma}_2\ \underline{L}' = \underline{\Delta}$, a diagonal matrix, we can clearly, by considering the transformation $\underline{y} = \underline{L}(\underline{x} - \underline{\mu}_1)$, confine the simulation without loss of generality to two distributions of the forms $N_p(\underline{0},\underline{I}_p)$ and $N_p(\underline{\nu},\underline{\Delta})$. In case we assume equality of $\underline{\Sigma}_1$ and $\underline{\Sigma}_2$ we take further, without loss of generality, $\underline{\Delta} = \underline{I}_p$.

Successive basic samples of sizes N_1 and N_2 and test samples of sizes n_1 and n_2 were generated for populations π_1 and π_2. From theoretical consideration it is known that the substantial differences between the eight estimators is small for large basic samples; hence we have concentrated attention on basic samples which are small in relation to the parameter dimension. A single simulation situation is characterised by a choice of values for the parameter set

$$(p, \underline{\nu}, \underline{\Delta}, N_1, N_2, n_1, n_2).$$

Table 1 provides details of the simulation parameter values under consideration here. The total number of combinations of parameter values was 702 with 351 for $\underline{\Sigma}_1 = \underline{\Sigma}_2$ and 351 for $\underline{\Sigma}_1 \neq \underline{\Sigma}_2$.

Table 1

Parameter values used in the comparison of the estimators

	p, $\underline{\nu}$, $\underline{\Delta}$ configurations
Equal covariance matrices	
$\underline{\nu} = \underline{J}$ a vector of unit elements	for $p = 2,3,\ldots,10$
$\underline{\nu} = (1,0,\ldots,0)'$	for $p = 2,3,\ldots,10$
$\underline{\nu} = (2,0,\ldots,0)'$	for $p = 2,3,\ldots,10$

Unequal covariance matrices

$$\left.\begin{array}{ll} \underline{\nu} = \underline{J}, & \underline{\Delta} = 4\ \underline{I}_p \\ \underline{\nu} = (1,0,\ldots,0)', & \underline{\Delta} = 4\ \underline{I}_p \\ \underline{\nu} = (2,0,\ldots,0)', & \underline{\Delta} = 16\ \underline{I}_p \end{array}\right\} \quad \text{for } p = 2,3,\ldots,10$$

Sizes of basic (N) and test (n) samples

p	(N_1,N_2)				(n_1,n_2)
2	$N_1=5$,	$N_2=5,6,\ldots,11$;	$N_2=5$,	$N_1=6,7,\ldots,11$	(15,15)
3	$N_1=6$,	$N_2=6,7,\ldots,12$;	$N_2=6$,	$N_1=7,8,\ldots,12$	
4	$N_1=7$,	$N_2=7,8,\ldots,13$;	$N_2=7$,	$N_1=8,9,\ldots,13$	
5	$N_1=8$,	$N_2=8,9,\ldots,14$;	$N_2=8$,	$N_1=9,10,\ldots,14$	
6	$N_1=9$,	$N_2=9,10,\ldots,15$;	$N_2=9$,	$N_1=10,11,\ldots,15$	
7	$N_1=10$,	$N_2=10,11,\ldots,16$;	$N_2=10$,	$N_1=11,12,\ldots,16$	
8	$N_1=11$,	$N_2=11,12,\ldots,17$;	$N_2=11$,	$N_1=12,13,\ldots,17$	
9	$N_1=12$,	$N_2=12,13,\ldots,18$;	$N_2=12$,	$N_1=13,14,\ldots,18$	
10	$N_1=13$,	$N_2=13,14,\ldots,19$;	$N_2=13$,	$N_1=14,15,\ldots,19$	

The measure of performance of estimator i that we adopt for the basic samples and for the test samples is a mean absolute deviation from the true discriminant function.

The eight estimators are pairwise compared for each simulation situation, both for basic samples and for test samples. The mean absolute deviation averaged over all combinations of parameter values are given in Table 2. The percentages of observations misassigned, averaged over all combinations of parameter values, are given in Table 3. The letters e and u refer to the assumptions of equal or unequal covariance matrices respectively in the construction of the estimators. Although it is the performance on the test samples that is most relevant to a comparison of the different estimators, estimation of discriminant function for observations in the basic samples may be needed if the original identification is subject to doubt or amendment. The differences of percentages of misassignments between the basic samples and the test samples are evidently essential in practice.

In the case of small samples for simulated situations with $\underline{\Sigma}_1 = \underline{\Sigma}_2$ and $\underline{\Sigma}_1 \neq \underline{\Sigma}_2$ we see from Tables 2 and 3 that:

1. The estimators $\hat{e}_4$ and $\hat{e}_2$ are superior to all other estimators and are themselves comparable.
2. The estimators $\hat{e}_1$, $\hat{e}_3$ and $\hat{u}_4$ are comparable and slightly inferior to $\hat{e}_4$ and $\hat{e}_2$.
3. The estimators $\hat{u}_1$, $\hat{u}_3$ and $\hat{u}_2$ are far inferior to the other five estimators and are themselves comparable.
4. With respect to percentage of misassignment there is little difference between the eight estimators.

Our general conclusions can therefore be summarized as follows:

(i) In the case of small samples the estimators $\hat{e}_4$ and $\hat{e}_2$ yield considerably better results than to the six other estimators, even for the situation involving unequal covariance matrices. These estimators are themselves comparable.

(ii) A substantial reduction in the mean absolute deviation of the usual estimators is achieved by adjustments for bias.

Table 2

Mean absolute deviation of the estimators averaged over all simulated models

Simulation type	Nature of sample	Estimator							
		$\hat{e}_1$	$\hat{e}_2$	$\hat{e}_3$	$\hat{e}_4$	$\hat{u}_1$	$\hat{u}_2$	$\hat{u}_3$	$\hat{u}_4$
$\underline{\Sigma}_1 = \underline{\Sigma}_2$	Basic	2.42	1.61	2.44	1.53	33.17	5.28	33.29	4.14
	Test	2.48	1.68	2.48	1.54	33.90	5.36	34.17	3.38
$\underline{\Sigma}_1 \neq \underline{\Sigma}_2$	Basic	3.07	2.85	3.07	2.83	129.26	16.58	129.37	9.00
	Test	3.38	2.97	3.38	2.88	90.61	13.01	90.82	8.67

Table 3

Percentage misallocated averaged over all simulated models

Simulation type	Nature of sample	Estimator							
		$\hat{e}_1$	$\hat{e}_2$	$\hat{e}_3$	$\hat{e}_4$	$\hat{u}_1$	$\hat{u}_2$	$\hat{u}_3$	$\hat{u}_4$
$\underline{\Sigma}_1 = \underline{\Sigma}_2$	Basic	13	13	14	13	5	35	5	6
	Test	29	29	29	29	38	43	38	38
$\underline{\Sigma}_1 \neq \underline{\Sigma}_2$	Basic	16	18	17	16	3	18	3	6
	Test	35	35	35	35	30	28	31	29

Appendix

Evaluation of the expected value of the estimator $\hat{u}_1(\underline{x})$

The evaluation of $E[\hat{u}_1(\underline{x})]$ may be most conveniently obtained by taking conditional and then unconditional expectations, viz.,

$$E[\hat{u}_1(\underline{x})] = \frac{1}{2}\sum_{i=1}^{2}(-1)^i(N_i-1)E_{\underline{\bar{x}}_i}\Big\{E_{\underline{A}_i^{-1}}[((\underline{x}-\underline{\bar{x}}_i)'\underline{A}_i^{-1}(\underline{x}-\underline{\bar{x}}_i)|\underline{\bar{x}}_i) - - (N_i-1)^{-1}\ell n|\underline{A}_i^{-1}|]\Big\} + \frac{p}{2}\ell n\frac{N_1-1}{N_2-1},$$

where $\underline{A}_i = (N_i-1)\underline{S}_i$, $i = 1,2$.

This requires the evaluation of terms such as $E[\ell n|\underline{B}|]$ where $B\sim W(\underline{\Omega},\nu)$.

Let the $p\times p$ matrix $\underline{B}$ be distributed as $W(\underline{\Omega},\nu)$.

Then the characteristic function of $\ell n|\underline{B}|$ is

$$\varphi(t) = E[|\underline{B}|^{it}] = |2\Omega|^{it}\prod_{j=1}^{p}\frac{\Gamma[\frac{1}{2}(\nu+2it+1-j)]}{\Gamma[\frac{1}{2}(\nu+1-j)]}$$

and

$$\varphi'(t) = i\varphi(t)\left\{\sum_{j=1}^{p}\Psi[\tfrac{1}{2}(\nu+2it+1-j)]+\ell n|2\underline{\Omega}|\right\}.$$

Hence we obtain

$$E[\ell n|\underline{B}|] = -i\varphi'(0) = \sum_{j=1}^{p}\Psi[\tfrac{1}{2}(\nu+1-j)]+\ell n|2\underline{\Omega}|.$$

Since $\underline{A}_i\sim W(\underline{\Sigma}_i,N_i-1)$, we obtain

$$E_{\underline{A}_i^{-1}}[\ell n|\underline{A}_i^{-1}|] = \sum_{k=1}^{p}\Psi[\tfrac{1}{2}(N_i-k)]+\ell n|\underline{\Sigma}_i|+p\,\ell n\,2,$$

and

$$E_{\underline{A}_i^{-1}}[(\underline{x}-\overline{\underline{x}}_i)'\underline{A}_i^{-1}(\underline{x}-\overline{\underline{x}}_i)|\overline{\underline{x}}_i] = (N_i-p-2)^{-1}(\underline{x}-\overline{\underline{x}}_i)'\underline{\Sigma}_i^{-1}(\underline{x}-\overline{\underline{x}}_i),$$

$i = 1,2$.

In order to evaluate of $E_{\overline{\underline{x}}_i}[(\underline{x}-\overline{\underline{x}}_i)'\underline{\Sigma}_i^{-1}(\underline{x}-\overline{\underline{x}}_i)]$ we observe that if the p-vector $\underline{X}$ is distributed as $N(\underline{\delta},\underline{\Omega})$, then $Y = (\underline{X}-\underline{a})'\underline{\Omega}^{-1}(\underline{X}-\underline{a})$ is distributed as $\chi^2(\beta;p)$, where $\beta = (\underline{\delta}-\underline{a})'\underline{\Omega}^{-1}(\underline{\delta}-\underline{a})$ and $\chi^2(\beta;p)$ denotes the random variable having a noncentral χ^2-distribution with p degrees of freedom and noncentrality parameter β.

Moreover, $E(Y) = E[\chi^2(\beta;p)] = \beta+p$.

Hence

$$E_{\overline{\underline{x}}_i}[(\underline{x}-\overline{\underline{x}}_i)'\underline{\Sigma}_i^{-1}(\underline{x}-\overline{\underline{x}}_i)] = \Delta_i^2(\underline{x}) + pN_i^{-1}, \quad i=1,2$$

and (3) holds.

Evaluation of the estimator $\hat{u}_3(\underline{x})$

The joint density function for $(\overline{\underline{x}}_i,\underline{A}_i)$ is

$$f(\overline{\underline{x}}_i,\underline{A}_i \mid \underline{\mu}_i,\underline{\Sigma}_i^{-1}) \propto |\underline{A}_i|^{(N_i-p-2)/2} |\underline{\Sigma}_i|^{-N_i/2} \times$$

$$\times \exp\left\{-\frac{1}{2}\,\mathrm{tr}\,\underline{\Sigma}_i^{-1}[\underline{A}_i+N_i(\overline{\underline{x}}_i-\underline{\mu}_i)(\overline{\underline{x}}_i-\underline{\mu}_i)']\right\}$$

which, upon combining with the assumed joint prior density (7); yields

$$t(\underline{\mu}_i,\underline{\Sigma}_i^{-1} \mid \overline{\underline{x}}_i,\underline{A}_i) \propto |\underline{\Sigma}_i|^{-(N_i-p-1)/2} \times$$

$$\times \exp\left\{-\frac{1}{2}\,\mathrm{tr}\,\underline{\Sigma}_i^{-1}[\underline{A}_i+N_i(\overline{\underline{x}}_i-\underline{\mu}_i)(\overline{\underline{x}}_i-\underline{\mu}_i)']\right\}$$

as the joint posterior density for $\underline{\mu}_i$ and $\underline{\Sigma}_i^{-1}$, i=1,2.

Hence, conditional on $\underline{\Sigma}_i^{-1}$, $\underline{\mu}_i \sim N(\underline{\bar{x}}_i, N_i^{-1}\,\underline{\Sigma}_i)$, i=1,2. Also, $\underline{\Sigma}_i^{-1} \sim W[(N_i-1)^{-1}\underline{S}_i^{-1},\ N_i-1]$, i.e., $\underline{\Sigma}_i^{-1}$ has marginally a Wishart distribution, i = 1,2.

Now, the evaluation of $\hat{u}_3(\underline{x})$ may be most conveniently obtained by taking conditional and then unconditional expectations, viz.,

$$\hat{u}_3(\underline{x}) = E[u(\underline{x})] =$$

$$= \frac{1}{2}\sum_{i=1}^{2}(-1)^i E_{\underline{\Sigma}_i^{-1}}\left\{E_{\mu_i}[\Delta_i^2(\underline{x})\,|\,\underline{\Sigma}_i^{-1}] - \ln|\underline{\Sigma}_i^{-1}|\right\}.$$

Since $[N_i\Delta_i^2(x)\,|\,\underline{x}] \sim \chi^2(\beta_i;p)$, where

$$\beta_i = N_i(\underline{x}-\underline{\bar{x}}_i)'\underline{\Sigma}_i^{-1}(\underline{x}-\underline{\bar{x}}_i) \text{ and } E[\chi^2(\beta_i;p)] = \beta_i+p,$$

we obtain

$$E_{\underline{\mu}_i}[\Delta_i^2(\underline{x})\,|\,\underline{\Sigma}_i^{-1}] = N_i^{-1}(\beta_i+p),\ i = 1,2.$$

Hence

$$E_{\underline{\Sigma}_i^{-1}}[N_i^{-1}(\beta_i+p) - \ln|\underline{\Sigma}_i^{-1}|] = D_i^2(x) - \ln|\underline{S}_i^{-1}| -$$

$$- \sum_{k=1}^{p}\left\{\Psi[\tfrac{1}{2}(N_i-k)] + \ln[2(N_i-1)^{-1}]\right\},\ i=1,2$$

and (8) holds.

REFERENCES

ABRAMOWITZ, M. and STEGUN, I.A. (eds) (1965). Handbook of Mathematical Functions. New York. Dover Publications.

ANDERSON, T.W. (1958). An Introduction to Multivariate Statistical Analysis. New York, Wiley.

CORNISH, E.A. (1954). The multivariate t-distribution associated with a set of normal sample deviates, Australian Journal of Physics, 7, 531-542.

GEISSER, S. (1966). Predictive discrimination. In Multivariate Analysis (P.R. Krishnaih, ed.), pp. 149-163. New York. Academic Press.

JEFFREYS, H. (1961). Theory of Probability, 3rd ed. Oxford University Press.

Miroslav Krzyśko
Uniwersitet im. Adama Mickiewicza
Instytut Matematyczny
ul. Matejki 48/49
PL-60-769 Poznań

Proc. of the 4th Pannonian Symp. on Math. Stat., Bad Tatzmannsdorf, Austria 1983
W. Grossmann, G. Pflug, I. Vincze, W. Wertz, eds

APPROXIMATION OF STOCHASTIC PROCESSES BY PIECEWISE POLYNOMIALS

MARGIT LÉNÁRD

Debrecen

1. INTRODUCTION

In this paper we present a spline-interpolational method for stochastic processes having L^p - differentiability properties. In [3] we studied an algorithm for time series approximation by spline functions. Here we extend those results for L^p - approximation of the trajectories of stochastic processes having L^p - differentiability properties. It means, that instead of L^∞ - approximation we prove L^p - approximation by an extremely simple spline process, which depends on the increments of the original process corresponding to a given subdivision of the parameter interval. Further, it is proved, that if the increments of the original process are replaced by the increments of a sample process, close enough to the original one in the L^p - sense, then the method provides the same approximation along the trajectories.

Here we use the term "spline" in a somewhat different meaning than the classical one (see e.g. [1]): it is a piecewise polynomial function which is continuously differentiable. It provides the same accuracy as the classical one, and its coefficients can easily be determined - without solving a system of equations.

In the sequel we shall use the notions of L^p - continuity and L^p - differentiability of a stochastic process. Let $(\Omega,\mathcal{A},\mathcal{P})$ be a probability space, $p \geq 1$ and $\xi_t:\Omega \to \mathbb{R}$ $(a\leq t\leq b)$ a stochastic process for which $\xi_t \in L^p(\Omega)$ for all t in $[a,b]$. We say that ξ_t is L^p - continuous, L^p - differentiable or L^p - continuously differentiable, if the L^p - valued function $t \to \xi_t$ is continuous, differentiable or continuously differentiable, respectively.

The L^p - derivative of ξ_t will be denoted by ξ_t' if there may be no confusion. If ξ_t is L^p - continuous or L^p - continuously differentiable, then $\omega_p(h;\xi)$ and $\omega_{1,p}(h;\xi)$ will denote the modulus of continuity of the L^p - valued function $t \to \xi_t$ and $t \to \xi_t'$, respectively.

2. CONSTRUCTION OF THE SPLINE PROCESS

Let $(\Omega,\mathcal{A},\mathcal{P})$ be a probability space, $[a,b]$ a real interval and $\xi_t:\Omega \to \mathbb{R}$ an L^p - continuously differentiable stochastic process $(a \leq t \leq b)$. Fix an equidistant subdivision

$a = t_o < t_1 < \ldots < t_n = b$ of $[a,b]$ where $t_k - t_{k-1} = h$, and define the spline process S_t for $a \leq t \leq b$ as follows:

$$(1) \quad S_t = S_{t,k} \qquad t_{k-1} \leq t \leq t_k,$$

where

$$(2) \quad S_{t,1} = \xi_{t_1} + \frac{1}{h}(\xi_{t_1} - \xi_{t_o})(t - t_1)$$

and for k = 2, ..., n

(3) $S_{t,k} = \xi_{t_k} + \frac{1}{h}(\xi_{t_k} - \xi_{t_{k-1}})(t-t_k) - \frac{1}{h^2}(\xi_{t_k} - 2\xi_{t_{k-1}} + \xi_{t_{k-2}})$

$$\left[(t-t_k)^2 + \frac{1}{h}(t-t_k)^3 \right].$$

THEOREM 2.1 Let ξ_t be an L^p- continuously differentiable process $(a \leq t \leq b)$. Then, for the spline process S_t defined by (1) - (3) we have:

$$||\xi_t - S_t||_p = \sigma(h) \qquad (a \leq t \leq b)$$

PROOF In the sequel we shall use the Taylor-formula with integral remainder term for L^p-valued functions (see [2] Vol. 1. Ch. 8. pp. 191).

As the L^p-valued function $t \to \xi_t$ is continuously differentiable, we can apply the Taylor-formula and we get

$$\xi_t = \xi_{t_k} + \left(\int_0^1 \xi'_{t_k + \zeta(t-t_k)}\, d\zeta\right)(t-t_k)$$

whenever

$t_{k-1} \leq t \leq t_k$ $(k=1,\ldots,n)$. Further let

$$\xi^*_{t,k} = \xi_{t_k} + \xi'_{t_k}(t-t_k),$$

then we have for $t_{k-1} \leq t \leq t_k$

$$||\xi_t - \xi^*_{t,k}||_p \leq || \int_0^1 \xi_{t_k+\zeta(t-t_k)} d\zeta - \xi'_{t_k}||_p \cdot h =$$

$$= ||\int_0^1 (\xi'_{t_k+\zeta(t-t_k)} - \xi'_{t_k})\, d\zeta \,||_p\, h \leq$$

$$\leq \int_0^1 ||\xi'_{t_k+\zeta(t-t_k)} - \xi'_{t_k}||_p \, d\zeta \cdot h \leq \omega_{1,p}(h;\xi)\, h$$

and

$$||\xi^*_{t,k} - S_{t,k}||_p \leq ||\xi'_{t_k} - \frac{1}{h}(\xi_{t_k} - \xi_{t_{k-1}})||_p \cdot h +$$

$$+ 2\, ||\xi_{t_k} - 2\,\xi_{t_{k-1}} + \xi_{t_{k-2}}||_p .$$

Using the Taylor-formula again we get the estimate

$$||\xi'_{t_k} - \frac{1}{h}(\xi_{t_k} - \xi_{t_{k-1}})||_p = ||\xi'_{t_k} - \int_0^1 \xi'_{t_k+\zeta h} d\zeta \,||_p \leq$$

$$\leq \int_0^1 ||\xi'_{t_k} - \xi'_{t_k+\zeta h}||_p d\,\zeta < \omega_{1,p}\,(h;\xi)$$

and similarly

$$||\xi_{t_k} - 2\xi_{t_{k-1}} + \xi_{t_{k-2}}||_p \leq h\; \omega_{1,p}(h;\xi) .$$

Finally we obtain

$$||\xi_t - S_t||_p \leq ||\xi_t - \xi^*_{t,k}||_p + ||\xi^*_{t,k} - S_{t,k}||_p \leq$$

$$\leq h\omega_{1,p}(h;\xi) + 3\; h\omega_{1,p}(h;\xi) = 4\; h\; \omega_{1,p}(h;\xi) .$$

REMARK 2.2 By a similar computation as above we have the following estimate for the derivatives:

$$||\xi_t' - S_t' || < 7\ \omega_{1,p}(h;\xi)$$

for all t in $[a,b]$.

3. STOCHASTIC PROPERTIES OF THE SPLINE PROCESS

As a consequence of Theorem 2.1 we obtain that our spline process provides an "asymptotically unbiased estimate" for L^1- continuously differentiable stochastic processes. More precisely, the following theorem is valid:

THEOREM 3.1 *Let the process* ξ_t *be* L^1*-continuously differentiable and* S_t *be defined as above. Then for the expectations we have:*

$$E(\xi_t) - E(S_t) = \sigma(h).$$

In the next theorem we show that stationarity in the wide sense of ξ_t implies that S_t is asymptotically stationary in the wide sense.

THEOREM 3.2 *Let the process* ξ_t *be stationary in the wide sense and* L^2*-continuously differentiable. Then the process* S_t *defined above satisfies*

$$E(S_t\ S_{t+s}) - E(S_t)E(S_{t+s}) = \phi(s) + \sigma(h)$$

for all t,s *with* $a \leq t, t+s \leq b$, *where* ϕ *is some function.*

PROOF Let t,s be arbitrary with $a \leq t \cdot t+s \leq b$. As L^2 - continuous differentiability implies L^1-continuous differentiability, we have by Theorem 2.1

$$||S_t - \xi_t||_2 = \sigma(h), \; ||S_t - \xi_t||_1 = \sigma(h),$$

$$||S_{t+s}-\xi_{t+s}||_2 = \sigma(h), \; ||S_{t+s}-\xi_{t+s}||_1 = \sigma(h),$$

and by assumption

$$E(\xi_t \xi_{t+s}) - E(\xi_t)E(\xi_{t+s}) = \phi(s)$$

where ϕ is a function. Then, by the Cauchy-Schwarz inequality it follows that

$$|E((S_t-\xi_t)S_{t+s})| \leq ||\xi_t - S_t||_2 ||S_{t+s}||_2$$

$$|E((S_{t+s}-\xi_{t+s})\,\xi_t)| \leq ||S_{t+s}-\xi_{t+s}||_2 ||\xi_t||_2$$

and hence

$$E((S_t-\xi_t)S_{t+s}) = \sigma(h), \; E((S_{t+s}-\xi_{t+s})\xi_t) = \sigma(h).$$

Finally, we have

$$E(S_t S_{t+s}) - E(S_t)E(S_{t+s}) = E((S_t-\xi_t)S_{t+s}) + E((S_{t+s}-\xi_{t+s})\xi_t) +$$

$$+ E(\xi_t\xi_{t+s}) - E(\xi_t)E(\xi_{t+s}) + E(\xi_{t+s}-S_{t+s})E(\xi_t) + E(\xi_t - S_t)E(S_{t+s}),$$

and by the previous relations our statement follows.

THEOREM 3.3 *Let* ξ_t *be an* L^p*-continuously differentiable process and* S_t *the spline process, constructed as above. Let* $\{\eta_{i,1}, \ldots, \eta_{i,m}\}$ *be a sample for the variable* ξ_{t_i} *and* $\tau_i = F_i(\eta_{i,1}, \ldots, \eta_{i,m})$ *a statistic* $(i=1,\ldots,n)$.
If $||\tau_i - \xi_{t_i}||_p < \varepsilon$ *for* $i=1,\ldots, n$, *then for the spline process* $\tilde{S}_t$ *constructed by the statistics* τ_i *we have*

$$||\tilde{S}_t - \xi_t||_p = \sigma(h) + 11\,\varepsilon.$$

PROOF Let $t_{k-1} \leq t \leq t_k$ $(k=1,\ldots,n)$, then

$$||S_t - \tilde{S}_t||_p = ||S_{t,k} - \tilde{S}_{t,k}||_p < ||\xi_{t_k} - \tau_k||_p +$$

$$+ ||(\xi_{t_k} - \xi_{t_{k-1}}) - (\tau_k - \tau_{k-1})||_p +$$

$$+2||(\xi_{t_k} - 2\xi_{t_{k-1}} + \xi_{t_{k-2}}) - (\tau_k - 2\tau_{k-1} + \tau_{k-2})|| < 11\varepsilon,$$

and

$$||\tilde{S}_t - \xi_t||_p < ||\tilde{S}_t - S_t||_p + ||S_t - \xi_t||_p$$

$$\leq 11\varepsilon + \sigma(h)$$

by Theorem 2.1

REMARK 3.4 This theorem can be applied by taking a sample for each ξ_{t_k}, constructing some statistics, for example the arithmetic means of the respective sample elements, and constructing the spline process supported by these statistics. Then some stochastic properties of the original process are reflected on the approximating spline process. We remark further, that by the linearity of our construction, linear relations between different processes are preserved. The same is true concerning Theorem 3.3 if we use linear statistics.

REFERENCES

AHLBERG, J. H. - NILSON, E. N. - WALSH, J. H.: The Theory of Splines and their Applications, Academic Press, New-York, 1967.

DIEUDONNÉE, J.: Grundzüge der modernen Analysis, VEB Deutscher Verlag der Wissenschaften, Berlin, 1972.

LÉNÁRD, M.: Approximation and short time prediction of economic time series by spline functions, Studia Sci. Math. Hung. 17 (1982), 235-241.

Margit Lénárd
Kossuth Lajos University
Department of Mathematics
H-4010 Debrecen, Pf. 12.

Proc. of the 4th Pannonian Symp. on Math. Stat., Bad Tatzmannsdorf, Austria 1983
W. Grossmann, G. Pflug, I. Vincze, W. Wertz, eds

THE SECRETARY PROBLEM WITH HESITATING CANDIDATES

TAMÁS F. MÓRI

Budapest

ABSTRACT

The classical best choice problem is investigated in the case when one can make more than 1 choices and the individuals can refuse to be chosen. Two versions of the problem are solved by the usual backward induction technique combined with the differential equation method.

1. INTRODUCTION

If the most popular optimal stopping problem were put to the vote, it would result with a high probability in the so-called secretary (or best choice) problem. In the classical drafting it runs as follows: n rankable candidates applying for a job appear in a random order at the secretary's office. The secretary is ordered to employ the best one. This problem does not seem to be easy, because at each arrival the secretary has to decide immediately whether the candidate is to be accepted or rejected, later recall is impossible. Besides he can't use information other than the relative ranks with respect to the preceding applicants. As we all know it, the position of the secretary is not hopeless, not even if the number of applicants tends to infinity. He has only to let the first n/e arrivals go by, then to select the first one better than all previous ones.

In this way the probability of success is kept bounded away from 0, it converges to $1/e = 0.3678\ldots$.

This nice problem can be traced back to Cayley, who considered the "full information" version of it more than a hundred years ago (Cayley (1875)). Analogous questions were investigated and answered by a great number of probabilists, especially in the last two or three decades when the general theory of optimal stopping began to flourish. Without endeavouring to completeness we mention some of them.

In 1964 Chow, Moriguti, Robbins and Samuels solved the problem of minimizing the expected rank of the candidate chosen. They showed that the expected rank remains bounded for increasing n (it converges to 3.869...). Later Rubin and Samuels (1977) proved that this surprising assertion remains true under a further, very severe constraint: only one individual at a time can be remembered, i.e. the secretary has a memory not larger than length one.

In 1966 a very interesting paper appeared by Gilbert and Mosteller. Among other generalizations they considered the case when the secretary is allowed to choose more than one candidates (but, of course, a limited number of them). The optimal stopping rule they obtained was given by a sequence of threshold indices $n \geq n_1 \geq n_2 \geq \ldots$ depending only on n, but not on the number of choices. In case of k choices the ith choice should fall on the first leader (= a person of relative rank 1) after the time n_{k+1-i} and after the (i-1)st selection, of course. That is, we let again the first n_k candidates go by, then choose the first leader. If he arrives earlier than n_{k-1}, we should wait before the second choice etc., the last choice should not be made before $n_1 \approx \frac{n}{e}$.

Another exciting generalization is due to Presman and Sonin (1972). They allowed the number of candidates to be a random variable with known distribution.

In 1973 Mucci presented a new and plausible method for solving complicated recursions which often appear when carrying out the standard backward induction technique of the theory of optimal stopping (Mucci (1973a), (1973b)).

We can also refer to many excellent handbooks, e.g. Dynkin and Yushkevich (1967), Shiryayev (1969), DeGroot (1970) or Chow, Robbins and Siegmund (1971).

2. RESULTS FOR IMPERFECT SELECTION

In the present paper we consider the classical secretary problem with an additional assumption. Our candidates hesitate about accepting the job: the ith one (in the order of arrivals) gives a positive answer with probability p_i. In other words it can happen that the selection of the chosen individual does not meet with success. Two versions of the problem will be investigated.

A. First we suppose the probabilities to be equal: $p_1 = p_2 = \ldots = p_n = p$. We want to select the best of all applicants. To do this we can make k choices. If a candidate refuses the invitation, we may have another choice instead of him. (This is the situation when one has to select the best member of a population and the selection itself may damage the individuals.)

B. In the above problem the probability of success is clearly less than p. It would be reasonable to aim at selecting the best candidate of those ones who are ready to accept the post. This seems to be hard unless we know the willingness of each person before selecting him. In the second case this additional supposition is made while the equality of p_i's is no more required (this case occurs when the individuals can be damaged during the comparing examination).

In both cases we deal with the limit properties of the optimal strategy and the success probability as the number of candidates tends to infinity (k remains fixed).

<u>Theorem 1</u>. (Case A)

a. The optimal stopping rule is similar to the one obtained by Gilbert and Mosteller in the special case p=1. That is, there exists a sequence of threshold indices

$$n \geq n_1 \geq n_2 \geq \ldots \geq 0$$

with the following property: the ith choice should fall on the first leader arriving after the first n_{k+1-i} candidates and after the (i-1)st choice. This sequence does not depend on k.

b. The limits $x_i = \lim\limits_{n\to\infty} n_i/n$ do exist and satisfy the following recursion:

$$x_1^{1-p}=p, \quad x_k^{1-p} = - \sum_{j=1}^{k-1} x_j^{1-p} \frac{(p \log x_j)^{k-j}}{(k-j)!} \quad \text{if} \quad 0<p<1 \qquad (1)$$

and

$$\log x_1 = -1, \ \log x_k = -1 - \sum_{j=1}^{k-1} \frac{(\log x_j)^{k+1-j}}{(k+1-j)!} \text{ if } p=1, \qquad (2)$$

or they are given in a more compact form by the following identity:

$$\sum_{j=1}^{\infty} x_j^{1-p+pt} t^{j-1} \equiv p \quad \text{for every } t, \ |t|<1. \qquad (3)$$

c. The limit as $n\to\infty$ of the success probability is

$$x_1 + x_2 + \ldots + x_k.$$

Letting t tend to 1 from below in (3) we obtain $\Sigma_j x_j = p$, thus the success probability can be arbitrarily close to the trivial upper bound p if the number of choices is large enough. It can be shown that the deviation from p tends to 0 at an exponential rate as the number of choices increases. On the other hand, part b. of Theorem 1 can not be used without estimating the rate of convergence, since in practice n is always finite. Regarding it one can prove that $n_i = nx_i + 0(1)$ (cf. Remark 2.1.(ii) of Mucci (1973a)).

Let us turn to the case B. If a person intends to refuse the invitation we know it at the moment when he appears, so we can simply ignore him as if he were unobservable. In the optimal selecting procedure the selection or rejection of an observable leader depends on three parameters:

- How many candidates have arrived before? (Suppose he is the ith one.)

- How many of them were observable? Suppose he is the ℓth observable one, $\ell \le i$.)

- How many selections were made before? (Suppose we are going to make our j^{th} choice.)

An "observable leader" is a person having relative rank 1 with respect to the preceding observable arrivals. The worth of this title depends mainly on ℓ, rather than on i. The role of parameter i is to show how many applicants are still left.

Let $p(x)$, $0<x\le 1$, be a positive continuous function and suppose that $p_i = p(\frac{i}{n})$. Introduce $\pi(x) = \int_x p(t)dt$ and let $\{x_j\}$ be the sequence of threshold numbers obtained in case A for p=1 (that special case was investigated by Gilbert and Mosteller).

<u>Theorem 2</u>. (Case B)

a. The optimal stopping rule is given by a sequence of twodimensional sets

$$\Gamma_j \subset \{(i,\ell)\in \mathbb{N}^2 : \ell \le i \le n\}$$

not depending on k. An observable leader should be selected if the corresponding pair of parameters (i,ℓ) belongs to Γ_{k+1-j}.

b. The sequence of sets $\frac{1}{n}\Gamma_k$ has a limit γ_k as $n\to\infty$, with respect to the Hausdorff metric. This limiting set is the following:

$$\gamma_k = \{(x,y)\in \mathbb{R}^2 : 0\le x\le 1, \frac{x_k}{1-x_k}\pi(x)\le y\le x\} \qquad (4)$$

(cf. Fig.1).

c. The limit as $n\to\infty$ of the success probability is $x_1 + x_2 + \ldots + x_k$.

The above theorem suggests a simple approximation for Γ_k

$$\tilde{\Gamma}_k = \{(i,\ell)\in\mathbb{N}^2 : \frac{x_k}{1-x_k} \sum_{r=i}^{n} p_r \le \ell \le i \le n\} \ .$$

The stopping rule based on the sequence $\tilde{\Gamma}_1$ $\tilde{\Gamma}_2$ ... is asymptotically optimal in the sense that the limit of the success probability is the same as in the optimal case. Another asymptotically optimal stopping rule can be given in terms of threshold indices: $n_k = n\pi^{-1}(\pi(0)(1-x_k))$ (however, the second order properties of this latter stopping rule are not so good as those of the first one).

The background behind these remarks will be enlightened by the following simple observation. At time i (at the ith arrival) we have seen about the $\sum_{r=1}^{i-1} p_r / \sum_{r=1}^{n} p_r$ part of all observable candidates, i.e. the random variable ℓ looses its random character in the limit. Consequently, for large n our problem B can be regarded as if it arose from the classical problem of Gilbert and Mosteller by a time scale transformation. The relation between the original relative time $x=i/n$ and the observable relative time x' is approximately

$$x' \sim \int_0^x p(t)dt \ / \int_0^1 p(t)dt = 1 - \frac{\pi(x)}{\pi(0)} \ .$$

The asymptotic optimality of the stopping rules given above turns on the fact that in both cases the condition of the jth choice is asymptotically equivalent to the condition $x' \ge x_{k+1-j}$.

3. PROOF OF THE THEOREMS

In both cases we apply backward induction and write down recursions. These recursions will be transformed into a system of differential equations, which can be solved directly in case A. In case B we obtain complicated twovariate partial differential equations, but a suitable substitution reduces the problem to the first case.

Proof of Theorem 1.

a. Let $v_j(i)$ denote the maximal probability of selecting the best candidate where maximum is taken over all stopping rules allowing j or less choices, none of them is to be made before the ith arrival. A recursive technique for generating v_j is the following: $v_0(i)=0$ and for $j\geq 1$

$$v_j(n)=\frac{p}{n},$$

$$v_j(i)=\frac{p}{i}\max\left(\frac{i}{n}+v_{j-1}(i+1),\ v_j(i+1)\right)+\left(1-\frac{p}{i}\right)v_j(i+1) = \tag{5}$$

$$=v_j(i+1)+\frac{p}{i}\left(\frac{i}{n}-v_j(i+1)+v_{j-1}(i+1)\right)^+.$$

Obviously $v_j(i)$ increases in j and decreases in i. We are interested in the success probability $v_k(1)$. The optimal stopping rule is also defined by this recursion: having j choices left, we select a leader arriving at time i if $\frac{i}{n}+v_{j-1}(i+1)>v_j(i+1)$ (see e.g. Chow, Robbins and Siegmund (1971)). Using (5) it is easy to show that $\frac{i}{n}+v_{j-1}(i+1)$ increases in i, hence there exists a unique threshold index n_j such that the (k+1-j)th choice should not be made before it. Starting from (5) one can prove $n_j\leq n_{j-1}$ by induction (this fact is obvious also by intuition: we can begin the selection earlier if we are allowed to choose more).

b. The limit behaviour of $v_k(i)$ can be examined via differential equations, as it was done by Mucci (1973a). Let us normalize $v_k(i)$ by introducing the function $f_k(x)$, $0\le x\le 1$. Set

$$f_k(0) = v_k(1),$$

$$f_k(\tfrac{i}{n}) = v_k(i) \qquad i=1,2,\dots,n,$$

then extend f_k by linear interpolation to the entire unit interval. Note that f_k also depends on n. Applying the standard Picard method (see e.g. Pontryagin (1965)) one can show that f_k converges to a differentiable function g_k as $n\to\infty$, further g_k satisfies

$$g_k(1) = 0, \tag{6}$$

$$-g_k'(x) = \frac{p}{x}(x + g_{k-1}(x) - g_k(x))^+ \qquad 0<x<1,$$

where $g_0(x)\equiv 0$.

Putting $\Delta_k(x)=g_k(x)-g_{k-1}(x)$ define x_k as the smallest solution of the equation $x=\Delta_k(x)$. Then $1>x_1>x_2>\dots>0$ and $x>x_k$ implies $x>\Delta_k(x)$, while for $x\le x_k$ $\Delta_k(x)=x_k$ holds. Hence the limit of the success probability is

$$g_k(0)=g_k(x_k)=g_{k-1}(x_{k-1})+x_k=\dots=x_1+x_2+\dots+x_k .$$

We wish to determine these threshold numbers. In order to do it we rewrite (6) in the following way: for $x\ge x_k$

$$(x^{-p}g_k(x))'=x^{-p}g_k'(x)-px^{-p-1}g_k(x)=-px^{-p}-px^{-p-1}g_{k-1}(x),$$

hence for $p<1$ (we are ready to assume it, since the case $p=1$ has already been investigated by Gilbert and Mosteller) we obtain

$$g_k(x) = r(x^p - x) + px^p \int_x^1 t^{-p-1} g_{k-1}(t)dt$$

where $r = \frac{p}{1-p}$. Consequently

$$\Delta_1(x) = g_1(x) = r(x^p - x) \quad \text{for} \quad x \geq x_1 = p^{\frac{1}{1-p}} \tag{7}$$

and

$$\Delta_k(x) = px^p \int_x^1 t^{-p-1} \Delta_{k-1}(t)dt \quad \text{for} \quad x \geq x_{k-1}, \tag{8}$$

while for $x_k \leq x \leq x_{k-1}$

$$\Delta_k(x) = r(x^p - x) + px^p \int_x^{x_{k-1}} t^{-p-1} g_{k-1}(x_{k-1})dt +$$

$$+ x^p x_{k-1}^{-p} (g_k(x_{k-1}) - r(x_{k-1}^p - x_{k-1})) =$$

$$= -rx + x^p x_{k-1}^{-p} (\Delta_k(x_{k-1}) + rx_{k-1})$$

Using (7), (8) and (9) one can verify that for $x_j \leq x \leq x_{j-1}$, $j \leq k$ ($x_0 = 1$)

$$\Delta_k(x) = (-r)^{k+1-j} x + (-r)^k x^p \Big[-\sum_{i=0}^{k-1} \frac{(1-p)^i}{i!} (\log x)^i +$$

$$+ \sum_{\ell=1}^{j-1} (-r)^{-\ell+1} \left(\frac{x_\ell}{x_1}\right)^{1-p} \sum_{i=0}^{k-\ell} \frac{(1-p)^i}{i!} (\log x - \log x_\ell)^i \Big]. \tag{10}$$

The calculation is lengthy but straightforward, therefore we omit most details. We operate by induction in j and k.

For $j=1$, $k=1$ (10) coincides with (7).

For $j=1<k$ one can prove by induction in k, using (8).

For $1<j<k$ perform double induction, supposing the asser-

tion is proved for (j,k-1) and (j-1,k). Start from the equality

$$\Delta_k(x)=x^p x_{j-1}^{-p}\ \Delta_k(x_{j-1})+px^p \int_x^{x_{j-1}} t^{-p-1}\ \Delta_{k-1}(t)dt\ . \tag{11}$$

Apply case (j-1,k) to the first term and case (j,k-1) to the integrand.

Finally, for j=k we have to show that the right-hand sides of (9) and (10) are equal. Introduce

$$A_m = \sum_{i=1}^{m} \left(\frac{x_i}{x_1}\right)^{1-p} \frac{(p\ \log\ x_i)^{m-i}}{(m-i)!} \qquad m=1,2,...$$

and rewrite the right-hand side of (10) in terms of these expressions.

$$\begin{aligned}\Delta_k(x) &= (-r)^{k+j-1}x\ + \\ &+ (-r)^k x^p \sum_{\ell=0}^{k-1} \frac{(1-p)^\ell}{\ell!}(\log\ x)^\ell(-1+\sum_{m=1}^{k-\ell}(-r)^{-m+1}A_m) - \\ &-(-r)^k x^p \sum_{\ell=0}^{k-j}\frac{(1-p)^\ell}{\ell!}(\log\ x)^\ell \sum_{m=j}^{k-\ell}(-r)^{-m+1}\ \times \\ &\times \sum_{s=j}^{m}\left(\frac{x_s}{x_1}\right)^{1-p}\frac{(p\ \log\ x_s)^{m-s}}{(m-s)!}\end{aligned} \tag{12}$$

for $x_j \le x \le x_{j-1}$, $j \le k$.

Clearly $A_1=1$. Suppose that (10), or equivalently, (12) is proved for (j,j), j<k, further $A_2=A_3=...=A_{k-1}=0$. From (12) with j=k-1 it follows that

$$\Delta_k(x_{k-1}) = -rx_{k-1} - rx_{p-1}^p(A_k - \left(\frac{x_k}{x_1}\right)^{1-p}).$$

Substituting this into (9) one can easily derive (12) with j=k. Hence

$$\Delta_k(x_k) = x_k - rx_k^p A_k,$$

thus $A_k=0$ and the proof of (10) is completed.

From $A_k=0$ a simple rearrangement gives (1), while (3) follows from the identity

$$\sum_{m=1}^{\infty} x_1^{1-p} A_m t^{m-1} \equiv p.$$

So far we have dealt with the case p<1. Analogous computations can be carried out for p=1 resulting

$$g_k(x) = \sum_{\ell=1}^{j-1} x_\ell + x \sum_{\ell=0}^{k+1-j} \frac{(-\log x)^\ell}{\ell!} \sum_{m=j}^{k+1-\ell} \sum_{s=j}^{m} \frac{(\log x_s)^{m-s}}{(m-s)!} \qquad (13)$$

for $x_j \le x \le x_{j-1}$, $j \le k$, and $g_k(x) = x_1 + x_2 + \ldots + x_k$ for $x \le x_k$, where the sequence $\{x_j\}$ satisfies (2). This case can be treated also by limiting considerations:

$$pA_{k+1} = \sum_{i=1}^{k+1} x_i^{1-p} \frac{(p \log x_i)^{k+1-i}}{(k+1-i)!} =$$

$$= \sum_{i=1}^{k+1} \frac{(\log x_i)^{k+1-i}}{(k+1-i)!} + 0(1-p)$$

as $p \to 1$, from this (2) immediately follows.

Proof of Theorem 2.

Let $v_j(i,\ell)$ denote the maximal probability of selecting the best observable person in j trials, supposed we have seen i-1 candidates up to now, ℓ-1 of them were observable, and we haven't chosen yet. Further, let $Q(i,\ell)$ denote the probability that the candidate arriving in the ith place is the best observable person, supposed he is an observable leader and there were ℓ observable ones among the first i arrivals. Clearly

$$Q(i,\ell) = \sum_{r=0}^{n-i} \frac{\ell}{\ell+r} \sum_{H} \prod_{s\in H} p_s \prod_{s\in H^c} (1-p_s)$$

where the second summation runs over such subsets of $\{i+1,\dots,n\}$, which have exactly r elements, further $H^c=\{i+1,\dots,n\}\setminus H$. A more convenient expression for $Q(i,\ell)$ is

$$Q(i,\ell) = \ell\int_1^0 (1-t)^{\ell-1} \prod_{r=i+1}^{n} (1-p_r t)dt. \tag{14}$$

From this it follows that $Q(i,\ell)$ is increasing in i, and also in ℓ (it can be seen after integrating by parts).

The backward induction method leads to the following recursion: $v_0(i,\ell)=0$ and for $j\geq 1$

$$v_j(n,\ell)=p_n/\ell,\ 1\leq\ell\leq n,$$

$$\begin{aligned} v_j(i,\ell) &= \frac{p_i}{i} \max(Q(i,\ell)+v_{j-1}(i+1,\ell+1),\ v_j(i+1,\ell+1))+ \\ &\quad + p_i(1-\tfrac{1}{i})v_j(i+1,\ell+1)+(1-p_i)v_j(i+1,\ell) = \\ &= v_j(i+1,\ell) + p_i(v_j(i+1,\ell+1) - v_j(i+1,\ell)) + \\ &\quad + \frac{p_i}{i}(Q(i,\ell) + v_{j-1}(i+1,\ell+1) - v_j(i+1,\ell+1))^+ \end{aligned} \tag{15}$$

$1\leq\ell\leq i<n.$

Define

$$\Gamma_j = \{(i,\ell)\in\mathbb{N}^2:\ \ell\leq i\leq n, \text{ either } i<n \text{ and } Q(i,\ell) + v_{j-1}(i+1,\ell+1)\geq v_j(i+1,\ell+1), \text{ or } i=n\},$$

the optimal stopping rule is given by these sets in the way described in part a. of Theorem 2. In order to find the limit of $\frac{1}{n}\Gamma_k$ we normalize $v_k(i,\ell)$ again and write down the continuous analogue of difference equations (15). Thus let

$$f_k(x,y) = \lim_{n\to\infty} v_k(\lceil nx\rceil, \lceil ny\rceil) \qquad 0<y\le x\le 1$$

where $\lceil . \rceil$ stands for upper integer part, the existence of $f_k(x,y)$ and validity of the partial differential equation

$$\frac{\partial}{\partial x} f_k(x,y) + p(x)\frac{\partial}{\partial y} f_k(x,y) =$$

$$= -\frac{p(x)}{x}(q(x,y) + f_{k-1}(x,y) - f_k(x,y))^+, \quad (16)$$

$$f_k(1,y) = 0,$$

$$f_o(x,y) \equiv 0$$

can be proved by standard methods of the theory of differential equations. Here

$$q(x,y) = \lim_{n\to\infty} Q(\lceil nx\rceil, \lceil xy\rceil) = y\int_0^\infty \exp(-t(y+\pi(x)))dt =$$

$$= \frac{y}{y+\pi(x)} \quad .$$

The Hausdorff limit of $\frac{1}{n}\Gamma_k$ is

$$\gamma_k = \{(x,y)\in\mathbb{R}^2 \colon 0\le y\le x\le 1,\ q(x,y)+f_{k-1}(x,y)\ge f_k(x,y)\} \quad .$$

Define $g_k(z,w) = f_k(\pi^{-1}(z(1-w)), zw)$, i.e. substitute

$$z = y + \pi(x),$$

$$w = \frac{y}{y+\pi(x)} = q(x,y).$$

Then the domain $\{0 \le y \le x \le 1\}$ corresponds to $\{0 \le w \le 1,\ 0 \le z \le \frac{\pi(0)}{1-w}\}$, further $\frac{\partial x}{\partial w} = \frac{z}{p(x)}$, $\frac{\partial y}{\partial w} = z$, and (16) turns into

$$\frac{\partial}{\partial w} g_k(z,w) = -\frac{1}{w}\,(w+g_{k-1}(z,w) - g_k(z,w))^+ ,$$

$$g_k(z,1) = 0, \tag{17}$$

$$g_0(z,w) \equiv 0.$$

It can be shown by induction in k that $g_k(z,w)$ does not depend on z. Denoting $g_k(z,w)$ shortly by $g_k(w)$ we can observe that (17) concides with (6), hence g_k is given by (13). Hence $f_k(x,y) = g_k(q(x,y))$ and

$$\gamma_k = \{(w,z)\colon x_k \le w \le 1,\ 0 \le z \le \frac{\pi(0)}{1-w}\} =$$

$$= \{(x,y)\colon 0 \le x \le 1,\ \frac{x_k}{1-x_k}\,\pi(x) \le y \le x\} ,$$

which has to be proved.

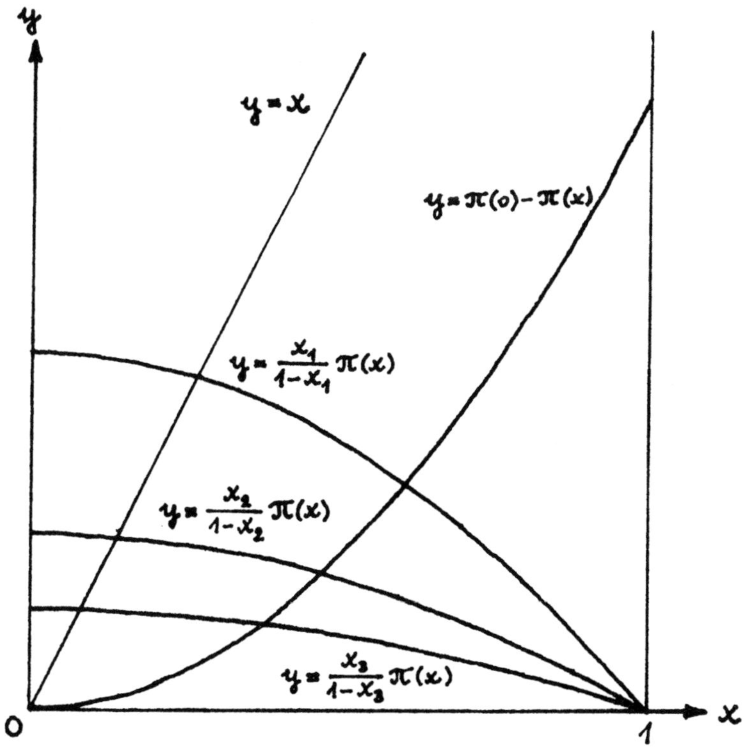

Figure 1

REFERENCES

CAYLEY, A. (1875) Mathematical questions and their solutions. Educational Times 22, 18-19.

CHOW, Y.S., MORIGUTI, S., ROBBINS, H. and SAMUELS, S.M. (1964) Optimum selection based on relative rank (the "Secretary Problem"). Israel J. Math. 2, 81-90.

CHOW, Y.S., ROBBINS, H. and SIEGMUND, D. (1971) Great Expectations: The Theory of Optimal Stopping. Houghton Mifflin, Boston.

COWAN, R. and ZABCZYK, J. (1976) A new version of the best choice problem. Bull. Acad. Polon. Sci. Sér. Math. Astronom. Phys. 24, 773-778.

DeGROOT, M.H. (1970) Optimal Statistical Decisions. McGraw-Hill, New York.

DYNKIN, E.B. and YUSHKEVICH, A.A. (1967) Markov Processes, Theorems and Problems (in Russian), Nauka, Moscow.

GARDNER, M. (1960) Mathematical games. Sci. Amer. 202, 150-156, 173-182.

GILBERT, J.P. and MOSTELLER, F. (1966) Recognizing the maximum of a sequence. J. Amer. Statist. Assoc. 61, 35-73.

GUSEIN-ZADE, S.M. (1966) The problem of choice and the optimal stopping rule for a sequence of independent trials (in Russian), Teor. Verojatnost. i Primenen. 11, 534-537.

MUCCI, A.G. (1973a) Differential equations and optimal choice problems. Ann. Statist. 1, 104-113.

MUCCI, A.G. (1973b) On a class of secretary problems. Ann. Probab. 1, 417-427.

NIKOLAEV, M.L. (1977) On a generalization of the problem of best choice (in Russian), Teor. Verojatnost. i Primenen. 22, 191-194.

PONTRYAGIN, L.S. (1965) Ordinary Differential Equations (in Russian), Nauka, Moscow.

PRESMAN, E.L. and SONIN, I.M. (1972) The best choice problem for a random number of objects (in Russian), Teor. Verojatnost. i Primenen. 17, 695-706.

RUBIN, H. and SAMUELS, S.M. (1977) The finite memory secretary problem. Ann. Probab. 5, 627-635.

SHIRYAYEV, A.N. (1969) Sequential Statistical Analysis (in Russian), Nauka, Moscow.

Tamás F. Móri
Department of Probability Theory
Loránd Eötvös University
Muzeum krt. 6-8.
H-1088. Budapest

Proc. of the 4th Pannonian Symp. on Math. Stat., Bad Tatzmannsdorf, Austria 1983
W. Grossmann, G. Pflug, I. Vincze, W. Wertz, eds

PROBABILITY DISTRIBUTIONS AND ASYMPTOTICS FOR SOME CHARACTERISTICS OF RANDOM MAPPINGS

LJUBEN MUTAFČIEV

Sofia

A large number of papers (see, for example, the survey papers [2] and [10]) are devoted to investigations of graphs with certain elements of randomness. Usually, the vertex set of a random graph is preliminary determined and the randomness is introduced by the hypothesis that each edge occurs with a given probability. Erdös and Rényi [5], [6] were the first who initiated the study of random **graphs** and proved interesting results about their structural properties as, for example, connectedness, the existence of a subgraph of specified type and so on. Solutions of several problems of this sort are described in details in the survey papers [2] and [10], where the recent developments in this field are also presented.

In this paper we focus our attention on a special type of random digraphs with n labelled vertices and n oriented arcs (n is a natural number). The vertex set of these graphs is $X_n =\{1,2,\dots,n\}$. Let M_n be the set of all single-valued mappings of the set X_n into itself. Clearly $|M_n|=n^n$. Each mapping $T\in M_n$ has a unique graph-representation G_T : for each $i,j\in X_n$, an arc goes from i to j iff $j=T(i)$. The graphs G_T, constructed in this way, consist of a number of components with exactly one cycle in each.

There are two equivalent points of view for defining the probability measure on the set M_n. We may consider $T(1), T(2), \ldots, T(n)$ as independent random variables with distributions $P\{T(i)=j\}=n^{-1}$, $j=1,\ldots,n$, so that the existence of an arc (i,j) in G_T is ruled by a probability, equal to n^{-1}. Thus, the occurence probability of a fixed G_T-construction, $T\in M_n$, is n^{-n}. This is equivalent to the assumption that each T is picked up at random from the set M_n (i.e. with probability $|M_n|^{-1}=n^{-n}$). In this way all numerical characteristics of T become random variables.

Fairly many papers (see the respective part of the book [16]) concern various characteristics and properties of random graphs G_T representing mappings T. Basic results are obtained in [11], [9], [17], [12], [3], [14]. In both papers [11], [9] the graphs G_T are interpreted as a description of acquaintance-type relations in the group of individuals X_n. In [8] G_T are suggested for models of epidemic processes: namely, X_n is considered as a population; a contagious disease is initially confined to a subset $A\subset X_n$, and it is then spread to other vertices (individuals) along arcs of G_T. Three versions of epidemic spread have been so far described. We shall restrict ourselves to the third version when the infection spreads in both arc-directions of G_T assuming that A is a fixed one-element set, say, $A=\{i_0\}$. Let ξ_n, η_n, and ζ_n be random variables, described as follows. Here ξ_n is the number of vertices in the path connecting i_0 with the nearest cyclic vertex (without counting the cyclic vertex); η_n is the number of vertices in the component containing i_0; ζ_n is the number of vertices in the cycle of the component containing i_0.

In Section 2 of this paper the exact form and the asymptotic, as $n\to\infty$, of the probability distribution of the random vector (ξ_n,η_n,ζ_n) are obtained. The one-dimensional marginal distributions and the distribution of the vector $(\xi_n+\zeta_n,\zeta_n)$ are known for a long time [9]. It is easy to see that η_n is the part of X_n, infected by i_o in the random graph-epidemic model under cosideration. Usually, ξ_n is said to be the height of the vertex i_o. Our main purpose in this paper is to determine the limiting behavior of η_n/n under some restrictions on the height ξ_n of i_o and on the cyclic size ζ_n. More exactly, the following theorem is proved in Section 3.

Theorem. Let $n\to\infty$ and $k=o(\sqrt{n})$; then

$$P\{\eta_n/n = y|\xi_n = k\} = P\{\eta_n/n = y|\zeta_n = k\}(1+o(1))= \frac{1+o(1)}{n\pi\sqrt{y(1-y)}} \quad (1)$$

where $0<y<1$.

Thus, it turns out that the density of the arcsine law is the limit of the last conditional probabilities. This theorem extends a previous result [13], obtained for $P\{\eta_n/n<x|\xi_n=0\}$, where $0<x<1$.

2. The Probability Distribution of the Random Vector (ξ_n,η_n,ζ_n)

A mapping $T\in M_n$ is said to be indecomposable if its graph-representation G_T has exactly one component. Let I_n denote the number of all indecomposable mappings. It is well-known [11] that

$$I_n = (n-1)! \sum_{k=0}^{n-1} \frac{n^k}{k!}, \quad n=1,2,\ldots . \quad (2)$$

Further, as usual, we put $(s)_\tau = s(s-1)\ldots(s-\tau+1)$ for every s and every natural number τ.

In this section we give the exact probability distribution of ξ_n, η_n, ζ_n, and its asymptotic, as $n\to\infty$.

Lemma 1. (a) Let n be fixed; then

$$P\{\xi_n=k,\eta_n=\ell,\zeta_n=m\}=n^{-n}\binom{n-1}{\ell-1}(n-\ell)^{n-\ell}\frac{m+k}{\ell}(\ell)_{m+k}\ell^{\ell-m-k-1};\qquad \ell=1,\ldots,n;\ m=1,\ldots,\ell;\ k=0,\ldots,\ell-m. \tag{3}$$

(b) Let $n\to\infty$; then the asymptotic density of the random vector $(\xi_n/\sqrt{n},\eta_n/n,\zeta_n/\sqrt{n})$ is equal to

$$\frac{(x+z)\exp\{-(x+z)^2/2y\}}{y\sqrt{2\pi y(1-y)}} \tag{4}$$

where $0<y<1$, $0<x<\infty$, $0<z<\infty$.

Proof. Notice first that for a given $\eta_n=\ell$ we may consider the component containing the fixed vertex i_0 as a labelled connected graph with ℓ vertices and ℓ arcs, picked at random from the set of all such graphs, i.e. from the set of all indecomposable mappings of $\{1,2,\ldots,\ell\}$ into itself. Let ξ_ℓ^* and ξ_ℓ^* be the number of vertices in the path from i_0 to the nearest cyclic vertex and the number of vertices in the cycle of a random indecomposable mapping, respectively. Thus, for the distribution of (ξ_n,ζ_n) conditional upon the event $\{\eta_n=\ell\}$, we get

$$P\{\xi_n=k,\zeta_n=m\mid\eta_n=\ell\}=P\{\xi_\ell^*=k,\zeta_\ell^*=m\};\qquad \ell=1,\ldots,n;\ m=1,\ldots,\ell;\ k=0,\ldots,\ell-m, \tag{5}$$

However the distribution of $(\xi_\ell^*,\zeta_\ell^*)$ is known to be [1]:

$$P\{\xi_\ell^*=k,\zeta_\ell^*=m\}=I_\ell^{-1}\,\frac{m+k}{\ell}(\ell)_{m+k}\,\ell^{\ell-m-k-1}; \qquad (6)$$

$$\ell=1,\ldots,n;\ m=1,\ldots,\ell;\ k=0,\ldots,\ell-m,$$

where the numbers I_ℓ are given by formula (2). On the other hand, it can be easily verified by simple combinatorial arguments that

$$P\{\xi_n=\ell\}=n^{-n}\binom{n-1}{\ell-1}(n-\ell)^{n-\ell}I_\ell,\quad \ell=1,\ldots,n. \qquad (7)$$

Combining relations (5) - (7) we get the distribution in (3).

In order to prove assertion (b) of the lemma we claim that

$$(\ell)_{m+k}\ell^{-m-k}=\exp\{-(m+k)^2/2\ell\}(1+O((m+k)^3/\ell^2)), \qquad (8)$$

as $\ell\to\infty$ and $m+k=o(\ell^{2/3})$. Indeed, the left-hand side of (8) can be represented in the form

$$(\ell)_{m+k}\ell^{-m-k}=\left(\frac{\ell^\ell}{\ell}\,e^{-\ell}\right)^{-1}\frac{\ell^{\ell-m-k}}{(\ell-m-k)!}\,e^{-\ell} \qquad (9)$$

The multipliers of the right-hand side of (9) are the probabilities that a Poisson-distributed random variable with parameter ℓ takes values ℓ and $\ell-m-k$, respectively. Now (8) follows by the local limit approximation of the Poisson distribution; as the parameter tends to infinity (see [7], Chapter 7, §4).

Replacing (8) in (3) and applying Stirling's formula, we obtain that

$$P\{\xi_n=k,\eta_n=\ell,\zeta_n=m\}$$

$$= \frac{\frac{m+k}{\sqrt{n}} : \frac{\ell}{n}}{n^2\sqrt{2\pi\frac{\ell}{n}(1-\frac{\ell}{n})}} \exp\left\{\left(\frac{m+k}{\sqrt{n}}\right)^2 : \frac{2\ell}{n}\right\}\left[1+O\left(\left(\frac{k+m}{\sqrt{n}}\right)^3 / \sqrt{n}\left(\frac{\ell}{n}\right)\right)\right],$$

as $n\to\infty$, $\ell=O(n)$, and $m+k=o(n^{2/3})$. This relation shows the asymptotic equivalence

$$P\left\{\frac{\xi n}{\sqrt{n}} = x, \frac{\xi n}{n} = y, \frac{\xi n}{n} = z\right\} = \frac{(x+z)\exp\{-(x+z)^2/2y\}}{n^2y\sqrt{2\pi y(1-y)}} (1+o(1))$$

for any positive numbers x, y, and z such that $x+z=o(n^{1/6})$ and $y<1$. Part (b) of lemma 1 is an immediate consequence of this asymptotic formula.

3. Conditional Probability Distributions of the Random Variable η_n/n : Convergence to the Arcsine Law

In order to prove the result of the theorem formulated in the introduction, we need the marginal distributions of the random vector (ξ_n,η_n,ζ_n). The following two lemmas describe the two-dimensional distributions.

Lemma 2. (a) Let n be fixed; then

$$P\left\{\xi_n = k,\eta_n = \ell\right\} = n^{-n}\binom{n-1}{\ell-1}(n-\ell)^{n-\ell}\left(1-\frac{k}{\ell}\right)(\ell)_k\ell^{\ell-k-1}; \qquad (10)$$

$$\ell=1,\ldots,n;\ k=o,\ldots,\ell-1;$$

$$P\left\{\eta_n = \ell, \zeta_n = m\right\} = n^{-n}\binom{n-1}{\ell-1}(n-\ell)^{n-\ell}(\ell)_m \ell^{\ell-m-1}; \qquad (11)$$

$$\ell=1,\ldots,n;\ m=1,\ldots,\ell.$$

(b) Let $n\to\infty$ and $k=o(n^{2/3})$; then

$$P\left\{\xi_n = k,\ \eta_n/n = y\right\} = P\left\{\eta_n/n = y,\ \zeta_n = k\right\}(1+o(1)) \qquad (12)$$

$$= \frac{\exp(-k^2/2ny)}{n\sqrt{n}\sqrt{2\pi y(1-y)}}(1+o(1))$$

where $0<y<1$.

Proof. Relations (10) and (11) follow after proper summations in (3) using the identities [1]:

$$\frac{1}{\ell}\sum_{m=1}^{\ell-k}(m+k)(\ell)_{m+k}\,\ell^{\ell-m-k-1} = (\ell)_k\,\ell^{\ell-k-1}\left(1-\frac{k}{\ell}\right),$$

$$\frac{1}{\ell}\sum_{k=0}^{\ell-m}(m+k)(\ell)_{m+k}\,\ell^{\ell-m-k-1} = (\ell)_m\,\ell^{\ell-m-1},$$

respectively. Now applying Stirling's formula and formula (8) for $m=0$ in (10) and (11), we get immediately (12).

Note. Of course, from lemma 2 it follows that the asymptotic densities of the random vectors $(\xi_n/\sqrt{n}, \eta_n/n)$ and $(\eta_n/n, \zeta_n/\sqrt{n})$ coincide, as $n\to\infty$, and are equal to $[\exp(-x^2/2y)]/\sqrt{2\pi y(1-y)}$, $0<x<\infty$, $0<y<1$, which agrees with expression (4).

Lemma 3. Let n be fixed; then

$$P\{\xi_n=k,\ \zeta_n=m\} = (n)_{m+k}\,n^{-m-k-1};\ k=0,\ldots,n-1;\ m=1,\ldots,n. \qquad (13)$$

Proof. Summing up with respect to ℓ in (3), we get

$$P\{\xi_n=k,\ \zeta_n=m\} = n^{-n}\sigma(n,m,k), \tag{14}$$

where

$$\sigma(n,m,k) = \sum_{\ell=m+k}^{n} \binom{n-1}{\ell-1}(n-\ell)^{n-\ell}\,\frac{m+k}{\ell}\,(\ell)_{m+k}\,\ell^{\ell-m-k-1} \tag{15}$$

On the other hand,

$$\begin{aligned}
\sigma(n,m,k) &= (m+k)\sum_{\nu=m+k-1} \frac{1}{1+\nu}\binom{n-1}{\nu}(n-1-\nu)^{n-1-\nu}(\nu)_{m+k-1}(\nu+1)^{\nu-m-k+1}\\
&= (m+k)\sum_{\nu=m+k-1}^{n-1}\left(\frac{1}{1+\nu}\right)(n-1-\nu)^{n-1-\nu}\left[\frac{d^{m+k-1}}{dx^{m+k-1}}(x+\nu)^{\nu}\right]\Bigg|_{x=1}\\
&= (m+k)\frac{d^{m+k-1}}{dx^{m+k-1}}\left[\sum_{\nu=m+k-1}^{n-1}\frac{1}{1+\nu}\binom{n-1}{\nu}(n-1-\nu)^{n-1-\mu}(x+\nu)^{\nu}\right]\Bigg|_{x=1}\\
&= \frac{m+k}{n}\frac{d^{m+k-1}}{dx^{m+k-1}}\left[\sum_{\mu=1}^{n}\binom{n}{\mu}(n-\mu)^{n-\mu}(x+\mu-1)^{\mu-1}\right]\Bigg|_{x=1}.
\end{aligned}$$

Applying Abel's identity

$$\sum_{k=0}^{n}\binom{n}{k}(x+k)^{k}(y+n-k)^{n-k} = \sum_{k=0}^{n}(n)_k(x+y+n)^{n-k}$$

(see [15], Chapter 1, §5) for the sum after the differential operator, we obtain the following expression for $\sigma(n,m,k)$:

$$\begin{aligned}
\sigma(n,m,k) &= \frac{m+k}{n}\frac{d^{m+k-1}}{dx^{m+k-1}}\left[\frac{(x+n-1)^n-n^n}{x-1}\right]\Bigg|_{x=1}\\
&= \frac{m+k}{n}(m+k-1)!\ \mathrm{Coeff}_{x^{m+k-1}}\left[\frac{(x+n-1)^n-n^n}{x-1}\right] \qquad (16)\\
&= \frac{(m+k)!}{n}\binom{n}{m+k}n^{n-m-k}.
\end{aligned}$$

A combination of (14) - (16) yields (13).

Note. It can be easily proved that

$$P\left\{\frac{\xi_n}{\sqrt{n}}=x,\ \frac{\zeta_n}{\sqrt{n}}=y,\right\}=\frac{1}{n}\,e^{-(x+y)^2/2}(1+o(1)),\quad n\to\infty,$$

for any positive numbers x and y satisfying $x+y=o(n^{1/6})$ (see the proof of lemma 1). In other words, $\exp\{-(x+y)^2/2\}$, $0<x<\infty$, $0<y<\infty$, is the asymptotic density of $(\xi_n/\sqrt{n},\zeta_n/\sqrt{n})$.

Let us consider now the one-dimensional marginal distributions of (ξ_n,η_n,ζ_n). The distribution of the random variable η_n is given by (7). The following lemma gives the distributions of ξ_n and ζ_n.

Lemma 4. (a) Let n be fixed; then

$$P\{\xi_n=k\}=P\{\zeta_n=k+1\}=\sum_{j=k+1}^{n}(n)_j\,n^{-j-1},\ k=0,\dots,n-1. \tag{17}$$

(b) Let $n\to\infty$ and $k=o(n^{2/3})$; then

$$P\{\xi_n=k\}=P\{\zeta_n=k+1\}=\frac{1}{\sqrt{n}}\,g_n(k)(1+o(1)), \tag{18}$$

where

$$g_n(k)=\sqrt{\frac{\pi}{2}}-\frac{1}{\sqrt{n}}\sum_{j=0}^{k}e^{-j^2/2n}\ . \tag{19}$$

Proof. Relations (17) follow immediately from (13). In order to prove (18), we use the following representation:

$$\begin{aligned}\sum_{j=0}^{n}(n)_j n^{-j}&=n\sum_{j=0}^{n}\binom{n}{j}j!\,n^{-j-1}=\sum_{j=0}^{n}\binom{n}{j}\int_0^\infty x^j e^{-nx}dx\\&=n\int_0^\infty(1+x)^n e^{-nx}dx\ .\end{aligned} \tag{20}$$

By standard asymptotic techniques (see [4], §4.4), for the last integral we get

$$n \int_0^{\infty}(1+x)^n e^{-nx}dx = \sqrt{\frac{\pi n}{2}}(1+o(1)), \quad n\to\infty. \tag{21}$$

Since

$$(n)_j n^{-j} = e^{-j^2/2n}(1+o(1)), \quad n\to\infty,$$

for each $j\le k$ and $k=o(n^{2/3})$, then from (20) and (21), we have

$$\sum_{j=k+1}^{n} (n)_j n^{-j-1} = \frac{1}{n}\left[\sum_{j=0}^{n}(n)_j n^{-j} - \sum_{j=0}^{k}(n)_j n^{-j}\right]$$

$$= \frac{1}{n}\left(\sqrt{\frac{\pi n}{2}} - \sqrt{n}\sum_{j=0}^{k} e^{-j^2/2n}\frac{1}{\sqrt{n}}\right)(1+o(1)) = \frac{1}{\sqrt{n}}\, g_n(k)(1+o(1)).$$

Taking into account (17), we obtain (18).

Note. Relations (18) and (19) show that the asymptotic densities of the random variables $\xi_n/\sqrt{n}$ and $\zeta_n/\sqrt{n}$ coincide and are equal to $\sqrt{2\pi}(1-\phi(x))$, $0<x<\infty$, where

$$\phi(x) = \frac{1}{\sqrt{2\pi}}\int_{-\infty}^{x} e^{-t^2/2}dt$$

The distribution of ζ_n and its limiting behavior are obtained in [9].

Now we are able to prove our main result for the conditional distributions of the random variable η_n/n.

Proof of the Theorem. Actually, the asymptotic given in (1) follows from (12), (18) and from the fact that $g_n(k)\to\sqrt{\frac{\pi}{2}}$, as $n\to\infty$ and $k=o(\sqrt{n})$.

Acknowledgements

I would like to thank the referee for the helpful comments. I am also grateful to Professor Wolfgang Wertz and Professor Bojan Penkov for improving the exposition.

REFERENCES

[1] G.N. Bagaev. Limit distributions of metric characteristics of a random indecomposable mapping. Combinatorial and asymptotic analysis (Krasnojarsk, 1977), pp. 55-61 (In Russian)

[2] B. Bollobás. Random graphs. Combinatorics, Proc. 8th Conf., Swansea 1981, Lond. Math. Soc. Lect. Notes Sor. 52 (1981), 80-102

[3] Y.D. Burtin. On a simple formula for random mappings and its applications. J. Appl. Probab., 17 (1980), 403-414

[4] N.G. De Bruijn. Asymptotic methods in analysis (North Holland, Amsterdam, 1958)

[5] P. Erdös, A. Rényi. On random graphs, I. Publ. Math. Debrecen, 6(1959), 290-297

[6] P. Erdös, A. Rényi. On the evolution of random graphs. Magyar Tud.Akad. Mat.Kutató Int. közl., 5(1960), 17-61

[7] W. Feller. An introduction to probability theory and its applications, Vol. I (John Wiley, New York, 1968, third ed.)

[8] I.B. Gertsbakh. Epidemic process on a random graphs: some preliminary results. J. Appl. Probab., 14(1977), 427-438

[9] B. Harris. Probability distributions related to random mappings Ann. Math. Statist., 31(1960), 1045 1062

[10] M. Karoński. A review of random graphs. J. Graph Theory, 6(1982), 349-389

[11] L. Katz. Probability of indecomposability of a random mapping function. Ann. Math. Statist., 26(1955), 512-517

[12] V.F. Kolchin. A problem of the allocation of particles in cells and random mappings. Theory Probability Appl., 21(1976), 48-63

[13] L.R. Mutafčiev. Limit distributions for the size of the random mapping's component containing a fixed element. Proc. 11th Spring Conf. of the Union of Bulg. Mathematicians, Sunny Beach, April 6-9, 1982. Sofia, BAN, 1982, pp. 404-411 (In Russian with English summary)

[14] B. Pittel. On distributions related to transitive closures of random finite mappings. Ann. Probab., 11(1983), 428-441

[15] J. Riordan. Combinatorial identities (John Wiley, New York, 1968)

[16] V.N. Sachkov. Probabilistic methods in combinatorial analysis (Nauka, Moscow, 1978) (In Russian)

[17] V.E. Stepanov. Limit distributions of certain characteristics of random mappings. Theory Probability Appl., 14(1969), 612-620

Ljuben Mutafčiev
Institute of Mathematics
Bulgarian Academy of Sciences
P. O. Box 373
BG-1090 Sofia

Proc. of the 4th Pannonian Symp. on Math. Stat.. Bad Tatzmannsdorf, Austria 1983
W. Grossmann, G. Pflug, I. Vincze, W. Wertz, eds

RECENT ADVANCES IN STATISTICAL APPLICATIONS OF COMMUTATION MATRICES

H. NEUDECKER

Amsterdam

ABSTRACT

Results of Magnus and Neudecker (1979) and Neudecker and Wansbeek (1983) will be generalized or extended. The assumption of normality will occasionally be dropped. Allowance will be made for singular variance matrices or non-zero expectations. The main result is an expression for the variance matrix of vec $X'AX$, $A'=A$, where $X'X$ follows a Wishart distribution. An illustration closes the paper.

1. INTRODUCTION

Magnus and Neudecker (1979) and Neudecker and Wansbeek (1983) presented a collection of statistical applications of commutation matrices.

At this juncture some new results will be communicated.

This opportunity will be seized to generalize or extend some earlier results. This will be of relevance for the derivation of the new results.

In the first place some results will be generalized.

This concern Theorems 4.3 and 4.4 of Magnus and Neudecker

Part of the research was done during the author's stay at the Social Sciences Faculty of the Hebrew University of Jerusalem. Their hospitality is thankfully acknowledged. Comments and help by Henk Don, Risto Heijmans and Tom Wansbeek are gratefully acknowledged.

(1979). It will be argued that it is imperative and possible to admit that the variance matrix $\bar{V}$ of the joint normal distribution of Theorem 4.3 is positive semi-definite. The theorem still holds under these circumstances, with the obvious exception of (v), about the rank of $D(x \otimes x)$.

Theorem 4.4 about (the variance matrix of) the Wishart distribution also holds if the variance matrix of the underlying normal distribution is positive semi-definite.

This has immediate consequences for the Neudecker and Wansbeek (1983) paper, which for some of its results relies on the earlier-mentioned theorems. These results can be accordingly rephrased. Fortunately not everywhere nonsingularity was explicitly assumed, so not everywhere alterations are needed. Other results can be generalized in a different way. Sometimes normality was unnecessarily assumed. This assumption will therefore be dropped if possible. Some other results can be amended so that they hold for possibly non-zero first moments. The main result of the paper will be an expression for the variance matrix of vec $X'AX$, $A'=A$, where X is an (n,p)-matrix, whose rows are stochastically independent $N_p(\mu,V)$ distributed vectors. This means that $X'X$ follows a Wishart $W_p(n,V,M)$ distribution, where $M' = (\mu,\dots,\mu) = s_n' \otimes \mu$. This result will be used in an illustrative way to derive the BQUE of V directly, i.e. not through maximum-likelihood methods.

2. Generalization of earlier results and extensions

We shall first present a generalized result concerning two jointly normally distributed vectors. The following theorem can be formulated.

<u>Theorem 1:</u>

Let the n-vector x and the m-vector y be jointly normally distributed.

Let $z' = (x':y')$ and assume that $E(z) = \mu = \begin{pmatrix}\mu_1\\ \mu_2\end{pmatrix}$ and $D(z) = \bar{V} = \begin{pmatrix}V & Q'\\ Q & R\end{pmatrix}$, where $\bar{V}$ is positive semi-definite, of rank k. Let further K_{mn} be the (mn,mn) commutation matrix as studied by Magnus and Neudecker (1979). Then

(i) $D(x\otimes y) = V\otimes R + V\otimes\mu_2\mu_2' + \mu_1\mu_1'\otimes R + K_{nm}(Q\otimes Q' + Q\otimes\mu_1\mu_2' + \mu_2\mu_1'\otimes Q')$

$= (Exx') \otimes (Eyy') + K_{nm}[(Eyx') \otimes (Exy')] - 2\mu_1\mu_1' \otimes \mu_2\mu_2'.$

(ii) $D(x\otimes x) = (I_{n^2}+K_{nn})(V\otimes V+V\otimes\mu\mu'+\mu\mu'\otimes V).$

<u>Proof</u> (Due to R. Heijmans):

Let $z = Zw + \mu$, where $Z = \begin{pmatrix}Z_1\\ Z_2\end{pmatrix}$ is an (n+m,k)-matrix of full column rank and $w\sim N_k(0,I)$.

Hence $\begin{pmatrix}V & Q'\\ Q & R\end{pmatrix} = \bar{V} = D(z) = ZZ' = \begin{pmatrix}Z_1Z_1' & Z_1Z_2'\\ Z_2Z_1' & Z_2Z_2'\end{pmatrix}$.

Further $\begin{pmatrix}x\\ y\end{pmatrix} = z = \begin{pmatrix}Z_1w+\mu_1\\ Z_2w+\mu_2\end{pmatrix}$.

Thus $x\otimes y = (Z_1w+\mu_1) \otimes (Z_2w+\mu_2)$.

From this point the proof goes as in the paper by Magnus and Neudecker (1979).

(ii) Let now $z' = (x':x')$. Hence $\mu_1=\mu_2=\mu$, $V=Q=Q'=R$ and $m=n$.

Then $D(x\otimes x) = (I_{n^2}+K_{nn})(V\otimes V+V\otimes\mu\mu'+\mu\mu'\otimes V)$, where $x\sim N_n(\mu,V)$.

The reader sees that it is <u>essential</u> to allow for singular $\bar{V}$ under (i), as $\bar{V} = \begin{pmatrix}V & V\\ V & V\end{pmatrix}$ is obviously singular.

This was pointed out by E. Neuwirth.

The proof of the following theorem is now straightforward.

Theorem 2:

Let S be Wishart $W_n(k,V,M)$ distributed with V positive semi-definite.

Then

$$D(\text{vec } S) = (I_{n^2} + K_{nn})[k(V\otimes V) + V\otimes M'M + M'M\otimes V].$$

Singularity of V does not introduce any complication, contrary to Khsirsagar's (1977) opinion.

We shall now give a theorem which incorporates non-zero first moments.

Theorem 3:

Let x be a $N_k(\mu,V)$ vector, where V is positive semi-definite.

Then

(i) $E(x\otimes x\otimes x\otimes x) = \text{vec}[K_{kk}(V\otimes V+V\otimes\mu\mu'+\mu\mu'\otimes V] + \text{vec}V\otimes\text{vec}V +$
$+ \text{vec}V\otimes\text{vec}\mu\mu'+\text{vec}\mu\mu'\otimes\text{vec}V+\text{vec}[(V+\mu\mu')\otimes(V+\mu\mu')].$

Let the n-vector x and the m-vector y be jointly normally distributed.

Let $z' = (x':y')$ and assume that

$E(z) = \mu = \begin{pmatrix}\mu_1\\ \mu_2\end{pmatrix}$. Then

(ii) $E(x\otimes x\otimes y\otimes y) = \text{vec}Exx'\otimes\text{vec}Eyy' +$
$+ (I_n\otimes K_{n,mm})[\text{vec}(Eyx'\otimes Exy') + \text{vec}Eyx'\otimes\text{vec}Exy'] -$
$- 2\text{vec}\mu_1\mu_1'\otimes\text{vec}\mu_2\mu_2';$

(iii) $E(x\otimes y\otimes x\otimes y) = \text{vec}(Exx'\otimes Eyy') + (I_{nm}\otimes K_{nm})[\text{vec}(Eyx'\otimes Exy') +$
$+ \text{vec}Eyx'\otimes\text{vec}Exy'] - 2\text{vec}(\mu_1\mu_1'\otimes\mu_2\mu_2');$

(iv) $E(x\otimes y\otimes y\otimes x) = (I_n\otimes K_{mm,n})(\text{vec}Exx'\otimes\text{vec}Eyy') + \text{vec}(Eyx'\otimes Exy')+$
$+ \text{vec}Eyx'\otimes\text{vec}Exy' - 2(I_n\otimes K_{mm,n})(\text{vec}\mu_1\mu_1'\otimes\text{vec}\mu_2\mu_2').$

Proof

(i) $E(x\otimes x\otimes x\otimes x) = E(\mathrm{vec}xx'\otimes \mathrm{vec}xx') = (I_k\otimes K_{kk}\otimes I_k)\mathrm{vec}E(xx'\otimes xx')$.

$D(x\otimes x) = E(xx'\otimes xx') - \mathrm{vec}Exx'(\mathrm{vec}Exx')' = E(xx'\otimes xx') -$

$- \mathrm{vec}(V+\mu\mu')\{\mathrm{vec}(V+\mu\mu')\}'$, hence

$(I_k\otimes K_{kk}\otimes I_k)\ \mathrm{vec}E(xx'\otimes xx') = (I_k\otimes K_{kk}\otimes I_k)\mathrm{vec}D(x\otimes x) +$

$+ (I_k\otimes K_{kk}\otimes I_k)[\mathrm{vec}(V+\mu\mu')\otimes \mathrm{vec}(V+\mu\mu')] =$

$= (I_k\otimes K_{kk}\otimes I_k)\mathrm{vec}[(I_{k^2}+K_{kk})(V\otimes V+V\otimes \mu\mu'+\mu\mu'\otimes V)] +$

$+ \mathrm{vec}[(V+\mu\mu')\otimes (V+\mu\mu')]$

$= \mathrm{vec}V\otimes \mathrm{vec}V + \mathrm{vec}V\otimes \mathrm{vec}\mu\mu' + \mathrm{vec}\mu\mu'\otimes \mathrm{vec}V +$

$+ \mathrm{vec}[K_{kk}(V\otimes V+V\otimes \mu\mu'+\mu\mu'\otimes V)] + \mathrm{vec}[(V+\mu\mu')\otimes (V+\mu\mu')]$.

(ii) $E(x\otimes x\otimes y\otimes y) = E(\mathrm{vec}xx'\otimes \mathrm{vec}yy') = (I_n\otimes K_{nm}\otimes I_m)\mathrm{vec}E(xx'\otimes yy')$.

$E(xx'\otimes yy') = E(x\otimes y)(x\otimes y)' = D(x\otimes y) + E(x\otimes y)E(x\otimes y)' =$

$= Exx'\otimes Eyy' + K_{nm}(Eyx'\otimes Exy') - 2\mu_1\mu_1'\otimes \mu_2\mu_2' +$

$+ \mathrm{vec}Eyx'(\mathrm{vec}Eyx')' = Exx'\otimes Eyy' + K_{nm}[Eyx'\otimes Exy'+\mathrm{vec}Exy'(\mathrm{vec}Eyx')']$

$- 2\mu_1\mu_1'\otimes \mu_2\mu_2'$, hence

$\mathrm{vec}E(xx'\otimes yy') = \mathrm{vec}(Exx'xEyy') + (I_{mn}\otimes K_{nm})[\mathrm{vec}(Eyx'\otimes Exy') +$

$+ \mathrm{vec}Eyx'\otimes \mathrm{vec}Exy'] - 2\mathrm{vec}(\mu_1\mu_1'\otimes \mu_2\mu_2')$, and

$(I_n\otimes K_{nm}\otimes I_m)\mathrm{vec}E(xx'\otimes yy') = \mathrm{vec}Exx'\otimes \mathrm{vec}Eyy' +$

$+ (I_n\otimes K_{n,mm})[\mathrm{vec}(Eyx'\otimes Exy') + \mathrm{vec}Eyx'\otimes \mathrm{vec}Exy'] -$

$- 2\mathrm{vec}\mu_1\mu_1'\otimes \mathrm{vec}\mu_2\mu_2'$.

(iii) $E(x\otimes y\otimes x\otimes y) = E(x\otimes \mathrm{vec}xy'\otimes y) = (I_n\otimes K_{mn}\otimes I_m)E(x\otimes \mathrm{vec}yx'\otimes y) =$

$= (I_n\otimes K_{mn}\otimes I_m)E(x\otimes x\otimes y\otimes y) = (I_n\otimes K_{mn}\otimes I_m)[\mathrm{vec}Exx'\otimes \mathrm{vec}Eyy'] +$

$+ (I_n\otimes K_{mn}\otimes I_m)(I_n\otimes K_{n,mm})[\mathrm{vec}(Eyx'\otimes Exy') + \mathrm{vec}Eyx'\otimes \mathrm{vec}Exy'] -$

$- 2(I_n\otimes K_{mn}\otimes I_m)(\mathrm{vec}\mu_1\mu_1'\otimes \mathrm{vec}\mu_2\mu_2') =$

$= \mathrm{vec}(Exx'\otimes Eyy') + (I_{nm}\otimes K_{nm})[\mathrm{vec}(Eyx'\otimes Exy') + \mathrm{vec}Eyx'\otimes \mathrm{vec}Exy'] -$

$- 2\mathrm{vec}(\mu_1\mu_1'\otimes \mu_2\mu_2')$.

(iv) $E(x\otimes y\otimes y\otimes x) = (I_n\otimes K_{mm,n})E(x\otimes x\otimes y\otimes y) =$

$= (I_n\otimes K_{mm,n})(\mathrm{vec}Exx'\otimes \mathrm{vec}Eyy') + \mathrm{vec}(Eyx'\otimes Exy') +$

$$- 2(I_n \otimes K_{mm,n})(\mathrm{vec}\mu_1\mu_1' \otimes \mathrm{vec}\mu_2\mu_2').$$

For proving these results we used: Theorem 3.1(i), Lemma 4.1(i) and (1.8) from Neudecker and Wansbeek (1983), and Theorem 1 and 2 of the present paper.

3. The variance matrix of vec X'AX

We shall nov prove the following result:

Theorem 4:

Consider n stochastically independent vectors $x_1,\ldots,x_n$, all of order p with $x_i \sim N_p(\mu,V)$ $(i=1,\ldots,n)$. Let s_n be a summation vector.

Consequently X'X, where $X'=(x_1,\ldots,x_n)$, will have a Wishart $W_p(n,V,M)$ distribution where $M'=(\mu,\ldots,\mu)=s_n'\otimes\mu$. Let A be a symmetric n-matrix. Then

(i) $E(X'AX) = M'AM + (\mathrm{tr}\ A)V = (s_n'As_n)\mu\mu' + (\mathrm{tr}\ A)V;$

(ii) $D(\mathrm{vec}X'AX) = (\mathrm{tr}\ A^2)(I_{p^2}+K_{pp})(V\otimes V) +$
$+ (s_n'A^2s_n)(I_{p^2}+K_{pp})(V\otimes V\mu\mu' + \mu\mu'\otimes V).$

Proof:

(i) Consider the more general case with $M'=(\mu_1,\ldots,\mu_n)$ and distinct V_i. Then

$$E(X'AX) = E(\sum_{ij} a_{ij}x_ix_j') = \sum_i a_{ii}E(x_ix_i') + \sum_{i\neq j} a_{ij}E(x_ix_j')$$

$$= \sum_i a_{ii}(V_i+\mu_i\mu_i') + \sum_{i\neq j} a_{ij}\mu_i\mu_j' = \sum_i a_{ii}V_i + \sum_{ij} a_{ij}\mu_i\mu_j'$$

$$= \sum_i a_{ii}V_i + M'AM.$$

In case of equal V_i's we get $E(X'AX) = (\mathrm{tr}\ A)V + M'AM$.

If further $M'=s_n'\otimes\mu$, we get $E(X'AX) = (\mathrm{tr}\ A)V + (s_n'As_n)\mu\mu'$, as $\sum_{ij} a_{ij} = s_n'As_n$.

Kshirsagar (1977) derived the result $E(X'AX) = mV + M'AM$ for symmetric idempotent A of rank m and distinct μ_i's.

(ii) Write $\text{vec}X'AX = (X'\otimes X')\text{vec}A = \text{vec}[I_{p^2}(X'\otimes X')\text{vec}A] =$
$= [(\text{vec}A)'\otimes I_{p^2}]\text{vec}(X'\otimes X')$. Hence $D(\text{vec}X'AX) =$
$= [(\text{vec}A)'\otimes I_{p^2}]D[\text{vec}(X'\otimes X')](\text{vec}A\otimes I_{p^2})$.

We use Theorem 1, and (3.2) from Neudecker and Wansbeek (1983) to write

$$D\ \text{vec}(X'\otimes X') = (I_{n^2p^2}+K_{nn}\otimes K_{pp})(I_n\otimes K_{np}\otimes I_p)[I_n\otimes V\otimes I_n\otimes V + I_n\otimes V\otimes \text{vec}M'(\text{vec}M')' + \text{vec}M'(\text{vec}M')'\otimes I_n\otimes V](I_n\otimes K_{pn}\otimes I_p).$$

We shall write the product of the last three terms as

$$(I_n\otimes K_{np}\otimes I_p)[I_n\otimes(V\otimes I_n)\otimes V + I_n\otimes(V\otimes s_ns_n')\otimes\mu\mu' + s_ns_n'\otimes(\mu\mu'\otimes I_n)\otimes V](I_n\otimes K_{pn}\otimes I_p) =$$
$$= I_n\otimes(I_n\otimes V)\otimes V + I_n\otimes(s_ns_n'\otimes V)\otimes\mu\mu' + s_ns_n'\otimes(I_n\otimes\mu\mu')\otimes V =$$
$$= I_{n^2}\otimes(V\otimes V) + (I_n\otimes s_ns_n')\otimes(V\otimes\mu\mu') + (s_ns_n'\otimes I_n)\otimes(\mu\mu'\otimes V).$$

Further $[(\text{vec}A)'\otimes I_{p^2}](I_{n^2p^2}+K_{nn}\otimes K_{pp}) = (\text{vec}A)'\otimes(I_{p^2}+K_{pp})$.

Hence

$$D(\text{vec}X'AX) =$$
$$= [(\text{vec}A)'\otimes(I_{p^2}+K_{pp})][I_{n^2}\otimes(V\otimes V) + (I_n\otimes s_ns_n')\otimes(V\otimes\mu\mu') + (s_ns_n'\otimes I_n)\otimes(\mu\mu'\otimes V](\text{vec}A\otimes I_{p^2}) =$$
$$= (\text{vec}A)'\text{vec}A\otimes(I_{p^2}+K_{pp})(V\otimes V) + (\text{vec}A)'\text{vec}s_ns_n'A\otimes(I_{p^2}+K_{pp})(V\otimes\mu\mu') + (\text{vec}A)'\text{vec}As_ns_n'\otimes(I_{p^2}+K_{pp})(\mu\mu'\otimes V)$$
$$= (\text{tr}A^2)(I_{p^2}+K_{pp})(V\otimes V) + (s_n'A^2s_n)(I_{p^2}+K_{pp})(V\otimes\mu\mu'+\mu\mu\otimes V).$$

This completes the proof.

Note:

It does not seem possible to treat the case of distinct μ_i, where $M' = (\mu_1,\ldots,\mu_n)$ by this procedure.

4. An illustration: the BQUE of V

We shall apply the expression for D(vecX'AX) to find the BQUE of V, where V is as defined in § 3. Of course, we know from maximum-likelihood theory that the BQUE is $\frac{1}{n-1}X'NX$, where $N = I_n - \frac{1}{n}s_n s_n'$. We shall, nevertheless, use an alternative approach which allows us to apply the new expression for D(vecX'AX). It follows from Theorem 4(i) that the unbiasedness conditions are

$$s_n'As_n = 0 \quad \text{and} \tag{1}$$

$$\mathrm{tr}A = 1 \tag{2}$$

Our aim will be to minimize z'D(vecX'AX)z for arbitrary z, subject to (1) and (2).

The two matrices $(I_{p^2}+K_{pp})(V\otimes V)$ and $(I_{p^2}+K_{pp})(V\otimes\mu\mu'+\mu\mu'\otimes V)$ are both positive semi-definite.

Let $\alpha = z'(I_{p^2}+K_{pp})(V\otimes V)z$ and $\beta = z'(I_p\ +K_{pp})(V\otimes\mu\mu'+\mu\mu'\otimes V)z$. Then our Lagrangean function will be

$$\varphi\{v(A)\} = \tfrac{1}{2}\alpha\mathrm{tr}A^2 + \tfrac{1}{2}\beta s'A^2s - \kappa s'As - \lambda(\mathrm{tr}A-1).$$

Minimizing A turns out to be $A = \frac{1}{n-1}(I_n - \frac{1}{n}s_n s_n')$.

(For the derivation see the Appendix.)

This result is, of course, well-known. In the absence of a more substantial illustration it is welcome.

Appendix

We shall minimize the scalar function

$$\varphi\{v(A)\} = \tfrac{1}{2}\alpha trA^2 + \tfrac{1}{2}\beta s'A^2s - \kappa s'As - \lambda(trA-1),$$

where $v(A)$ is the $\frac{1}{2}n(n+1)$-vector of the relevant elements of A, κ and λ being Lagrange multipliers.

We differentiate φ and get

$$d\varphi = \alpha trAdA - \kappa trss'dA - \lambda trdA + \beta trss'AdA =$$

$$= \{vec(\alpha A-\kappa ss'-\lambda I+\beta Ass')\}'Ddv(A) -$$

$$- (d\lambda)s'As - (d\kappa)(trA-1), \text{ hence}$$

$$0 = \frac{\partial\varphi}{\partial v(A)} = D'vec(\alpha A-\lambda ss'-\lambda I+\beta Ass') \qquad (3)$$

$$0 = s'As \qquad (1)$$

$$0 = trA - 1 \qquad (2)$$

are the first-order minimum conditions. D is the duplication matrix as defined by Magnus and Neudecker (1980).

As the function is convex there is no need to consider higher-order derivatives. We derive from (3)

$$2\alpha A - 2\kappa ss' + \beta Ass' + \beta ss'A - 2\lambda I - \alpha dgA + \kappa dgss' +$$

$$+ \lambda dgI - \beta dgAss' = 0, \text{ or}$$

$$2\alpha A - 2\kappa ss' + \beta Ass' + \beta ss'A - \alpha dgA - \beta dgAss' +$$

$$+ (\kappa-\lambda)I = 0, \text{ hence}$$

$$2\alpha trA - 2n\kappa + \beta s'As + \beta s'As - \alpha trA - \beta s'As + (\kappa-\lambda)n = 0$$

or

$$2\alpha - 2n\kappa - \alpha + (\kappa-\lambda)n = 0,$$

hence $\kappa + \lambda = \frac{1}{n}\alpha$. (4)

We used the obvious property: $trdgAss' = s'As$.

Further $2\alpha s'As - 2\kappa n^2 + \beta ns'As - \beta ns'As - \alpha s'(dgA)s -$

$- \beta s'(dgAss')s + n(\kappa-\lambda) = 0$, or

$- 2\kappa n^2 - \alpha + n(\kappa-\lambda) = 0$

as $s'(dgA)s = trA$

and $s'(dgAss')s = s'As$.

Hence $\alpha = n\kappa(1-2n)-n\lambda$. (5)

We can now solve for κ and λ:

$$\kappa = - \frac{\alpha}{n(n-1)} \tag{6}$$

$$\lambda = \frac{\alpha}{n-1} . \tag{7}$$

As the outcome for A should not be contingent on particular α and β, we can derive the following two equations now:

$$2A + \frac{2}{n(n-1)} ss' - dgA - \frac{n+1}{n(n-1)} I = 0 \tag{8}$$

and

$$Ass' + ss'A - dgAss' = 0. \tag{9}$$

The first equation yields

$$dgA = \frac{1}{n} I,$$

hence

$$A = \frac{1}{(n-1)} (I - \frac{1}{n} ss').$$

It is clear that the second equation is satisfied. Hence the optimal A has been found.

REFERENCES

Kshirsagar, A.M. (1977). Multivariate Analysis. Marcel Dekker, New York.

Magnus, J.R. and Neudecker, H. (1979). The commutation matrix: some properties and applications. The Annals of Statistics 7, 381-394.

Magnus, J.R. and Neudecker, H. (1980). The elimination matrix: some lemmas and applications. SIAM Journal on Algebraic and Discrete Methods. 1, 442-449.

Neudecker, H. and Wansbeek, T.J. (1983). Some results on commutation matrices, with statistical applications. The Canadian Journal of Statistics 11, 221-231.

Heinz NEUDECKER
Universiteit van Amsterdam
Instituut voor Actuariaat en Econometrie
Jodenbreetstraat 23.
NL-1011. NH AMSTERDAM

Proc. of the 4th Pannonian Symp. on Math. Stat., Bad Tatzmannsdorf, Austria 1983
W. Grossmann, G. Pflug, I. Vincze, W. Wertz, eds

QUASI-MONTE CARLO METHODS FOR GLOBAL OPTIMIZATION

HARALD NIEDERREITER

Vienna

1. Introduction

The underlying philosophy of a quasi-Monte Carlo method can be described briefly as follows. We recall first that a statistical Monte Carlo method is a procedure in numerical analysis that is based on a random sampling or simulation approach. If random sampling is replaced by a deterministic selection scheme that is well adjusted to the numerical problem at hand, we arrive at a quasi-Monte Carlo method. It can be argued that most practical implementations of statistical Monte Carlo methods are in fact quasi-Monte Carlo methods since the purportedly "random" samples are often generated in the computer by a deterministic algorithm. We refer to [14] for a survey of quasi-Monte Carlo methods.

In the area of global optimization of functions, statistical Monte Carlo methods are of interest in cases where the functions have a low degree of regularity, e.g. in the case of nondifferentiable functions in which the usual gradient methods fail. The Monte Carlo procedure that is used to approximate the global optimum of a function is that of random search. The idea of a random search method appeared in the statistical literature over forty years ago (see e.g. Hotelling [10]). More detailed studies were carried out later by Brooks [4], [5], Matyas [13], and Zieliński [24], among others. A discussion of Monte Carlo optimization techniques can be found in the book of Rubinstein [21, Ch. 7].

In this paper we will consider deterministic analogs of random search procedures for approximating the global optimum of a function. These deterministic procedures can therefore be thought of as quasi-Monte Carlo optimization methods or quasi--random search methods. In the analysis of quasi-Monte Carlo methods the following approach is typical: one first tries to establish an effective error bound in terms of a suitable quantity depending on the deterministically selected points, and then one seeks to find those sets of deterministic points that make this quantity as small as possible. For quasi-Monte Carlo optimization the relevant quantity is the so-called dispersion (see Section 2). In this study of quasi-Monte Carlo optimization methods we will, w.l.o.g., only consider the case where we are interested in the global supremum (or maximum) of a function.

In Section 2 we discuss the straightforward deterministic analog of a random search technique, namely a crude search method. Since such a method has several drawbacks, more refined methods are introduced in Section 3. Some theoretical results on the dispersion are presented in Section 4. The last section contains recommendations for the numerical practice regarding the choice of deterministic points to be used in quasi-random search procedures.

2. Crude search

Let X be a separable topological space and let f be a real-valued function on X that is bounded from above. We wish to approximate

$$M = \sup_{x \in X} f(x).$$

The following is a deterministic analog of a random search procedure. We select a sequence $x_1, x_2, \ldots$ of points of X and we calculate the numbers

$$m_N = \max_{1 \le n \le N} f(x_n) \quad \text{for} \quad N = 1, 2, \ldots .$$

Clearly we have

$$\lim_{N \to \infty} m_N = M \tag{1}$$

whenever f is continuous and the sequence $x_1, x_2, \ldots$ is dense in X. It is easy to construct examples which show that if one of these two conditions is not satisfied, then (1) fails in general.

The following stochastic model can be set up for this crude search procedure (compare e.g. with Zieliński [24]). Let μ be a probability measure on X and let $x_1, x_2, \ldots$ be independent μ-distributed samples from X. Then it is easily seen that

$$\lim_{N \to \infty} m_N = M \quad \text{almost surely,}$$

provided that f is continuous and μ is a Radon measure.

Crude search methods have been used in various settings; see Aird and Rice [1], Artobolevskiĭ et al. [2], Sobol' [22], and Sobol' and Statnikov [23]. A systematic theory including an error analysis was first developed by the author [15] (see also Niederreiter and McCurley [17]). For the purpose of error analysis, we let X be a totally bounded metric space with metric d, so that for every $\varepsilon>0$ we can cover X by finitely many balls of radius ε. The oscillation of f is measured by the modulus of continuity

$$\omega(t) = \sup_{\substack{x,y\in X \\ d(x,y)\le t}} |f(x) - f(y)| \quad \text{for } t\ge 0.$$

Furthermore, we define the dispersion of the points $x_1,\ldots,x_N \in X$ by

$$d_N = \sup_{x\in X} \min_{1\le n\le N} d(x,x_n).$$

Then the following error bound was established in [15]:

$$M - m_N \le \omega(d_N) \quad \text{for all } N\ge 1. \tag{2}$$

This error bound tends to 0 as $N\to\infty$ if f is uniformly continuous on X and $d_N\to 0$ as $N\to\infty$. The latter property is immediately seen to be equivalent to the denseness of the sequence $x_1,x_2,\ldots$. It is also easy to show that the bound in (2) is, in general, best possible.

It follows from (2) that suitable sets of deterministic points for quasi-random search methods are those with small dispersion. On the basis of this fact, numerical experiments with point sets of small dispersion were carried out by Niederreiter and McCurley [17] who took X to be the four-dimensional unit cube for the sake of concreteness. A detailed comparative analysis of the suitability of various low-dispersion point sets for quasi-Monte Carlo optimization is given in Niederreiter and Peart [18]. We shall return to the results of this analysis in Section 5.

3. Localization of search

The usefulness of the crude search method described in Section 2 is restricted by the fact that, in general, a very large number of function evaluations has to be carried out in order to get good approximations to M. Furthermore, most of the time is spent calculating function values at points that are far away from those $x \in X$ for which $f(x)$ is close to M. To remedy this situation, we introduce refinements of the crude search method.

As a first refinement we consider the method of "localization of search." Let X be again a totally bounded metric space with metric d. We denote by $B(a;\varepsilon)$ the closed ball with center $a \in X$ and radius ε. Since X is totally bounded, we have $X = B(b;r)$ for suitable $b \in X$ and $r>0$. For $B = B(a;\varepsilon)$ let g_B be a mapping from X onto B that should be easy to calculate and as close as possible to a contraction, i.e. to a mapping g with $d(g(x),g(y)) = c\, d(x,y)$ for some constant $c>0$ and all $x,y \in X$. Given N points $x_1,\ldots,x_N$ in X and a function f to be maximized, we set

$$m_N(B) = \max\left(f(a), \max_{1\le n\le N} f(g_B(x_n))\right) = f(x^*(B)), \tag{3}$$

where $x^*(B) \in B$ is one of the points at which f has been evaluated in this formula. For instance, if $X = B(b;r)$ is a closed ball in $\mathbb{R}^s$ with respect to a metric derived from a norm on $\mathbb{R}^s$, then for any closed ball $B = B(a;\varepsilon)$ in $\mathbb{R}^s$ with $a \in X$ we may take

$$g_B(x) = a + \frac{\varepsilon}{r}(x-b) \quad \text{for } x \in X.$$

Since $g_B(x)$ may fall outside of X in this case, we add the condition in (3) that the maximum over n is taken only over those $1\le n\le N$ with $g_B(x_n) \in X$.

Returning to the general case, we choose now a sequence $\varepsilon_1, \varepsilon_2, \ldots$ of positive numbers converging monotonically to 0 and define a sequence $B_0, B_1, \ldots$ of balls as follows:

$$B_0 = X, \; B_1 = B(x^*(B_0); \varepsilon_1),$$

$$B_{i+1} = \begin{cases} B(x^*(B_i); \varepsilon_{i+1}) & \text{if } \; m_N(B_i) > m_N(B_{i-1}), \\ B(x^*(B_{i-1}); \varepsilon_{i+1}) & \text{otherwise.} \end{cases} \qquad i = 1,2,\ldots$$

Then the sequence $m_N(B_0)$, $m_N(B_1), \ldots$ is nondecreasing and for suitably large i the value of $m_N(B_i)$ is taken as an approximation to M. In practice one often sets $\varepsilon_i = \varepsilon^i$ with a fixed ε, $0<\varepsilon<1$. The points $x_1, \ldots, x_N$ in X are then chosen in such a way that their dispersion satisfies $d_N < \varepsilon$.

If the sequence of ε_i converges too quickly to 0, it may happen that the sequence of $m_N(B_i)$ does not converge to M. The following procedure of "outer iteration" may be used to countervail that tendency. We first run through a small number of steps in the localization of search method, thus obtaining a point x_1^* at which the function f has the largest computed value. Then we start a second localization of search method with a ball $B(x_1^*; \delta)$ with δ larger than the radius of the ball searched last. In general, having arrived after i-1 outer iterations at the point x_{i-1}^*, we start the ith iteration with a ball $B(x_{i-1}^*; \lambda^{i-2}\delta)$, where $0<\lambda<1$. See [19] for a further discussion of this procedure.

If prior information on the function f is available, one could also employ adaptive methods to improve the performance of crude search methods. One idea is to use fewer points x_n in those regions where f varies slowly and more points in those regions where f varies quickly. Information on the oscillation of f can be obtained, for instance, from an initial crude search.

Another possibility is to adapt the metric to the behavior of f. For instance, if f is defined on a bounded interval X of $\mathbb{R}^s$, one may use the "weighted" metric

$$d(y,z) = \max_{1 \le j \le s} c_j |y_j - z_j|$$

for $y = (y_1,\ldots,y_s)$, $z = (z_1,\ldots,z_s) \in X$, where the positive weight c_j measures the oscillation of $f(y_1,\ldots,y_s)$ as a function of y_j. If f had bounded partial derivatives on X, we could use for c_j the supremum of $|f_{y_j}|$ over X. In general, these suprema will not exist or will not be available numerically, but we may use a difference scheme based on a coarse grid on X to get approximate values for the c_j. In order to obtain points in X with small dispersion relative to the "weighted" metric, we may use the mapping

$$g(y_1,\ldots,y_s) = (c_1 y_1,\ldots,c_s y_s) \text{ for } (y_1,\ldots,y_s) \in X,$$

then construct points $x_1,\ldots,x_N$ in g(X) with small dispersion relative to the standard metric with all $c_j = 1$ (compare with Section 4), and then take the points $g^{-1}(x_1),\ldots,g^{-1}(x_N)$ in X.

Numerical experiments using these refined methods have been performed by Niederreiter and Peart [19]; see Section 5 for a further description. We note also that stochastic models for refinements of the crude search method were set up and analyzed by Devroye [8] and Pintér [20].

4. Results on the dispersion

We have seen in Section 2 that the dispersion is the quantity that measures the suitability of point sets for quasi--random search methods. The theoretical study of the dispersion was already begun in [15]. A more extensive discussion of the dispersion, especially from the number-theoretic viewpoint, can be found in [16].

We consider a normalized situation in which $X = [0,1]^s = I^s$, the s-dimensional unit cube, and the metric is given by

$$d(y,z) = \max_{1 \le j \le s} |y_j - z_j|$$

for $y = (y_1,\ldots,y_s)$, $z = (z_1,\ldots,z_s) \in \mathbb{R}^s$. The minimum value of the dispersion d_N of N points in I^s is given below. We denote by $\lfloor t \rfloor$ the greatest integer $\le t$.

Theorem 1. For any N points in I^s we have

$$d_N \ge \frac{1}{2\lfloor N^{1/s} \rfloor} .$$

Furthermore, there always exist N points in I^s for which equality holds.

Proof. Let $x_1,\ldots,x_N$ be arbitrary points of I^s and let m be the positive integer with $m^s \le N < (m+1)^s$. If d_N is the dispersion of $x_1,\ldots,x_N$, then the balls B_n, $1 \le n \le N$, with center x_n and radius d_N cover I^s. Consider the $(m+1)^s$ points $(k_1/m,\ldots,k_s/m)$ with the k_j running independently through the set $\{0,1,\ldots,m\}$. By the pigeon-hole principle, some B_n contains two distinct points of this type, say p_1 and p_2. Then it follows that

$$\frac{1}{m} \le d(p_1,p_2) \le d(p_1,x_n) + d(x_n,p_2) \le 2d_N,$$

hence

$$d_N \ge \frac{1}{2m} = \frac{1}{2\lfloor N^{1/s} \rfloor} .$$

To prove the second part, let N be given and choose m as above. Consider the m^s points $(h_1/2m,\dots,h_s/2m)$ with the h_j running independently through the set $\{1,3,5,\dots,2m-1\}$, and then repeat some of these points to get a total of N points. For these N points it is easily seen that

$$d_N = \frac{1}{2m} = \frac{1}{2\lfloor N^{1/s}\rfloor} \; . \; \square$$

As a consequence, the minimal order of magnitude of the dispersion of N points in I^s is $N^{-1/s}$. But, as shown in [16], there also exist sequences of points in I^s with $d_N = O(N^{-1/s})$, where d_N is the dispersion of the first N points of the sequence. A more precise result can be deduced from the following lemma.

<u>Lemma 1.</u> If there exists a sequence of points in I with dispersion d_N satisfying $\overline{\lim\limits_{N\to\infty}}\, Nd_N = c<\infty$, then one can construct a sequence of points in I^s with dispersion $d_N^{(s)}$ satisfying $\overline{\lim\limits_{N\to\infty}}\, N^{1/s}d_N^{(s)}=c$.

<u>Proof.</u> Let $t_1,t_2,\dots$ be a sequence of points in I with $\overline{\lim\limits_{N\to\infty}}\, Nd_N=c$. Consider a sequence $x_1,x_2,\dots$ of points in I^s such that for any positive integer m the first m^s points of the sequence consist exactly of all $(u_1,\dots,u_s)$ with the u_j running independently through $t_1,\dots,t_m$. The precise order of the x_n is immaterial. Let $d_N^{(s)}$ be the dispersion of $x_1,\dots,x_N$, let $m^s\le N<(m+1)^s$, and let $y=(y_1,\dots,y_s)\in I^s$ be arbitrary. Then for each j, $1\le j\le s$, there exists n_j, $1\le n_j\le m$, with

$$|y_j-t_{n_j}| \le d_m.$$

The point $(t_{n_1},\dots,t_{n_s})$ is among the points $x_1,\dots,x_N$, thus $d_N^{(s)} \le d_m$. It follows that $N^{1/s}\, d_N^{(s)} < (m+1)d_m$, hence

$$\overline{\lim_{N\to\infty}}\, N^{1/s}\, d_N^{(s)} \le c.$$

On the other hand, for every positive integer m there exists $t \in I$ such that

$$\min_{1 \leq n \leq m} |t - t_n| = d_m. \tag{4}$$

Now consider $(t,\ldots,t) \in I^s$ and the first m^s points of the sequence $x_1, x_2, \ldots$. Then it follows from (4) that the distance from $(t,\ldots,t)$ to any of these points is at least d_m. Hence $d^{(s)}_{m^s} \geq d_m$, and so

$$\overline{\lim_{N\to\infty}}\, N^{1/s}\, d_N^{(s)} \geq c,$$

which completes the proof. □

<u>Theorem 2.</u> There exists a sequence of points in I^s with

$$\overline{\lim_{N\to\infty}}\, N^{1/s}\, d_N = \frac{1}{\log 4} = 0.721\ldots .$$

<u>Proof.</u> Let $t_1, t_2, \ldots$ be the sequence of points in I defined by

$$t_1 = 1,\ t_n = \left\{\frac{\log(2n-3)}{\log 2}\right\} \quad \text{for } n \geq 2,$$

where $\{t\} = t - \lfloor t \rfloor$ for real t. Then it was shown in [16] that this sequence satisfies the condition of Lemma 1 with $c = 1/\log 4$. The rest follows from Lemma 1. □

Theorem 2 is best possible for $s = 1$, since it was shown in [16] that for any sequence of points in I we have

$$\overline{\lim_{N\to\infty}}\, N d_N \geq \frac{1}{\log 4}\ .$$

For general s it follows of course from Theorem 1 that always

$$\underline{\lim_{N\to\infty}}\, N^{1/s}\, d_N \geq \frac{1}{2}\ .$$

In this connection it is also of interest to note that $N^{1/s}\, d_N$ is almost surely of the order of magnitude $(\log N)^{1/s}$ for sequences of points in I^s, as can be seen from the results of Deheuvels [7] on maximal spacings for multivariate order statistics.

Further examples of sequences of points in I^s satisfying $d_N = O(N^{-1/s})$ can be obtained, using the method in the proof of Lemma 1, by starting from one of the examples in [16] of a sequence of points in I satisfying the condition of Lemma 1. Another interesting class of examples for I^s is described in the sequel.

For real t, let $\{t\} = t - \lfloor t \rfloor$ and let $||t||$ denote the distance from t to the nearest integer. Let $\Theta_1,\dots,\Theta_s$ be real numbers satisfying the following diophantine condition: there exists a constant $c > 0$ such that

$$(\max_{1\le j\le s} |h_j|)^s\, ||h_1\Theta_1 + \dots + h_s\Theta_s|| > c \tag{5}$$

for all nonzero $(h_1,\dots,h_s)\in\mathbb{Z}^s$. For instance, this condition holds if $1,\Theta_1,\dots,\Theta_s$ form a basis of a real algebraic number field of degree s+1 over the rationals (see Cassels [6, pp. 79-80]).

<u>Theorem 3.</u> If $\Theta_1,\dots,\Theta_s$ satisfy (5), then for the sequence of points $(\{n\Theta_1\},\dots,\{n\Theta_s\})$, $n = 1,2,\dots,$ in I^s we have

$$d_N = O(N^{-1/s}).$$

<u>Proof.</u> It follows from (5) that given $H \ge 1$, there exists no nonzero $(h_1,\dots,h_s)\in \mathbb{Z}^s$ with

$$||h_1\Theta_1+\dots+h_s\Theta_s|| \le c\, H^{-s} \quad \text{and} \quad |h_j|\le H \quad \text{for } 1 \le j \le s.$$

An application of Theorem VIII on p. 84 of Cassels [6] shows that this property is equivalent to the following: there is a constant $c_1>0$ such that for any $M\geq 1$ and any $(y_1,\ldots,y_s)\in \mathbb{R}^s$ there exists an integer m with

$$||m\Theta_j-y_j||\leq c_1\ M^{-1/s} \quad \text{for} \quad 1\leq j\leq s \quad \text{and} \quad |m|\leq M. \tag{6}$$

To prove the theorem, we can assume $N\geq 3$. Put $M=\lfloor (N-1)/2\rfloor$ and let $u=(u_1,\ldots,u_s)\in I^s$ be arbitrary. It follows from (6) that there exist an integer m with $|m|\leq\lfloor (N-1)/2\rfloor$ and

$$||m\Theta_j+\left\lfloor\frac{N+1}{2}\right\rfloor\Theta_j-u_j||\leq c_1\left\lfloor\frac{N-1}{2}\right\rfloor^{-1/s} \quad \text{for } 1\leq j\leq s.$$

Putting $k=m+\lfloor (N+1)/2\rfloor$, we have $1\leq k\leq N$ and

$$||k\Theta_j-u_j||\leq CN^{-1/s} \quad \text{for } 1\leq j\leq s$$

with a constant $C>0$ independent of N and u. Now let $u\in I^s$ be such that

$$CN^{-1/s}\leq u_j\leq 1-CN^{-1/s} \quad \text{for} \quad 1\leq j\leq s. \tag{7}$$

Then

$$|\{k\Theta_j\}-u_j|=||k\Theta_j-u_j||\leq CN^{-1/s} \quad \text{for} \quad 1\leq j\leq s,$$

so that with $x_n=(\{n\Theta_1\},\ldots,\{n\Theta_s\})$ we get

$$\min_{1\leq n\leq N} d(u,x_n)\leq CN^{-1/s}.$$

If $u\in I^s$ does not satisfy (7), then we can find a point at a distance from u at most $CN^{-1/s}$ for which (7) holds. Consequently,

$$\min_{1\leq n\leq N} d(u,x_n)\leq 2CN^{-1/s}.$$

Altogether, we get $d_N\leq 2CN^{-1/s}$. □

5. Choice of deterministic points

We shall discuss various possibilities of choosing deterministic points for quasi-random search methods and the numerical experience accumulated so far. We consider again the situation in Section 4, namely $X = I^s$ and the metric d given there.

The criteria for choosing suitable point sets (or sequences) are the following: (i) small dispersion of the point set (or sequence); (ii) rapid calculation of the points. For the numerical practice it is also convenient to have the point set (or sequence) listed in such a way that every initial segment has a small dispersion as well.

It turns out that low-discrepancy sequences used for quasi-Monte Carlo integration (see [14, Sec. 3]) can also be employed in quasi-random search methods, since it was shown in [16] that we have the inequality

$$d_N \leq D_N^{1/s} \tag{8}$$

linking the dispersion d_N and the discrepancy D_N. For low--discrepancy sequences, D_N is of the order of magnitude $N^{-1}(\log N)^k$ with $k>0$ depending on s, so that (8) implies that d_N is, up to a logarithmic factor, of the order of magnitude $N^{-1/s}$.

A common method of obtaining low-discrepancy sequences is based on "good lattice points." Let m be a large integer and let $g=(g_1,\ldots,g_s)\in\mathbb{Z}^s$ be a lattice point. Consider the points

$$x_n = \left(\left\{\frac{n}{m}\, g_1\right\},\ldots,\left\{\frac{n}{m}\, g_s\right\}\right), \quad n=1,2,\ldots,m .$$

If the lattice point g is chosen judiciously, then the points $x_1,\ldots,x_m$ have a small discrepancy D_m (compare with [14, Sec. 4]) and thus a small dispersion d_m. Tables of good lattice points g are available in the literature (see e.g. Maisonneuve [12]).

Another possibility is to generate points from uniform pseudorandom numbers. Let $t_1,t_2,\ldots$ be a sequence of uniform pseudorandom numbers in I and define

$$x_n = (t_n,t_{n+1},\ldots,t_{n+s-1}) \quad \text{for} \quad n \geq 1.$$

Suitable t_n can be obtained, for example, by the linear congruential method. Appropriate parameters in this method that guarantee good distribution properties of the x_n can be found in the tables of Borosh and Niederreiter [3] and Knuth [11, Ch. 3]. This construction technique has the advantage that for each new point x_n only one new coordinate has to be calculated.

A sequence of points in I^s with very small discrepancy was recently constructed by Faure [9]. The calculation of this sequence is, however rather time-consuming. Some other methods of obtaining sequences with small dispersion are described in Niederreiter and Peart [18].

The numerical experiments reported in Niederreiter and Peart [18], [19] were based on a variety of methods of obtaining deterministic points, including those mentioned above. The data show clearly that if the real time necessary to achieve a prescribed accuracy is taken as a yardstick for performance, then the point sets generated from good lattice points and from linear congruential pseudorandom numbers are the most efficient. The numerical work in [19] demonstrates also that localization of search can be a very effective method for global optimization of multivariable functions.

REFERENCES

1 T. J. AIRD and J. R. RICE, Systematic search in high dimensional sets, SIAM J. Numer. Anal. 14, 296-312(1977).

2 I. I. ARTOBOLEVSKIĬ, M. D. GENKIN, V. K. GRINKEVIČ, I. M. SOBOL', and R. B. STATNIKOV, Optimization in the theory of machines by an LP-search, Dokl. Akad. Nauk SSSR 200, 1287-1290(1971). (Russian)

3 I. BOROSH and H. NIEDERREITER, Optimal multipliers for pseudo-random number generation by the linear congruential method, BIT 23, 65-74(1983).

4 S. H. BROOKS, A discussion of random methods for seeking maxima, Operations Research 6, 244-251(1958).

5 S. H. BROOKS, A comparison of maximum-seeking methods, Operations Research 7, 430-457(1959).

6 J. W. S. CASSELS, An Introduction to Diophantine Approximation, Cambridge Univ. Press, London, 1965.

7 P. DEHEUVELS, Strong bounds for multidimensional spacings, Z. Wahrscheinlichkeitsth. verw. Geb. 64, 411-424(1983).

8 L. P. DEVROYE, Progressive global random search of continuous functions, Math. Programming 15, 330-342(1978).

9 H. FAURE, Discrépance de suites associées à un système de numération (en dimension s), Acta Arith. 41, 337-351(1982).

10 H. HOTELLING, Experimental determination of the maximum of a function, Ann. Math. Statist. 12, 20-45(1941).

11 D. E. KNUTH, The Art of Computer Programming, Vol. 2: Seminumerical Algorithms, 2nd ed., Addison-Wesley, Reading, Mass., 1981.

12 D.MAISONNEUVE, Recherche et utilisation des "bons treillis". Programmation et résultats numériques, Applications of Number Theory to Numerical Analysis (S. K. Zaremba, ed.), pp. 121-201, Academic Press, New York, 1972.

13 J. MATYAS, Random optimization, Automation and Remote Control 26, 244-251(1965).

14 H. NIEDERREITER, Quasi-Monte Carlo methods and pseudo-random numbers, Bull. Amer. Math. Soc. 84, 957-1041(1978).

15 H. NIEDERREITER, A quasi-Monte Carlo method for the approximate computation of the extreme values of a function, Studies in Pure Mathematics (To the Memory of Paul Turán), pp. 523-529, Akadémiai Kiadó, Budapest, 1983. (Submitted in 1977).

16 H. NIEDERREITER, On a measure of denseness for sequences, Topics in Classical Number Theory (Budapest, 1981), North--Holland, Amsterdam, to appear.

17 H. NIEDERREITER and K. McCURLEY, Optimization of functions by quasi-random search methods, Computing 22, 119-123(1979).

18 H. NIEDERREITER and P. PEART, A comparative study of quasi-Monte Carlo methods for optimization of functions of several variables, Caribbean J. Math. 1, 27-44(1982).

19 H. NIEDERREITER and P. PEART, Localization of search in quasi-Monte Carlo methods for global optimization, preprint.

20 J. PINTÉR, On the convergence and computational efficiency of random search optimization, Methods of Operations Research 33, 347-362(1979).

21 R. Y. RUBINSTEIN, Simulation and the Monte Carlo Method, Wiley, New York, 1981.

22 I. M. SOBOL', On the systematic search in a hypercube, SIAM J. Numer. Anal. 16, 790-793(1979).

23 I. M. SOBOL' and R. B. STATNIKOV, LP-search and problems of optimal design, Problems of Random Search, Vol. 1, pp. 117-135, Izdat. "Zinatne," Riga, 1972. (Russian)

24 R. ZIELIŃSKI, On the Monte Carlo evaluation of the extremal value of a function, Algorytmy 2, no. 4, 7-13(1965).

Harald Niederreiter
Mathematical Institute
Austrian Academy of Sciences
Dr. Ignaz-Seipel-Platz 2
A-1010 Vienna
Austria

Proc. of the 4th Pannonian Symp. on Math. Stat., Bad Tatzmannsdorf, Austria 1983
W. Grossmann, G. Pflug, I. Vincze, W. Wertz, eds

EINGESCHRÄNKTE MONTE CARLO-VERFAHREN ZUR NUMERISCHEN INTEGRATION

ERICH NOVAK

Erlangen

ZUSAMMENFASSUNG

Zur näherungsweisen Berechnung von Integralen von Funktionen mehrerer Variablen sind i.a. sehr viele Stützstellen nötig, insbesondere dann, wenn die zu integrierende Funktion nur geringe Glattheitseigenschaften besitzt. So ist der Fehler $\left| \int_{[0,1]^s} f(x)\,d\lambda^s - \sum_{i=1}^{n} a_i f(x_i) \right|$ für einmal stetig differenzierbare Funktionen von der Größenordnung $n^{-\frac{1}{s}}$. Auch günstigste Quadraturverfahren können (wie erstmals BAHVALOV [1] gezeigt hat) diese Fehlerabschätzung nicht verbessern.
Dagegen kann man mit Hilfe von Monte Carlo-Verfahren (die Stützstellen x_i und die Gewichte a_i sind dann Zufallsvariablen) die Streuung der Zufallsvariablen $\sum_{i=1}^{n} a_i f(x_i)$ um das Integral $\int f(x)\,d\lambda^s$ von der Größenordnung $n^{-\frac{1}{s}-\frac{1}{2}}$ bestimmen, was eine erheblich schnellere Konvergenz bedeutet.

Nun sind aber Monte Carlo-Verfahren, die auf der Idee einer zufälligen Auswahl aus einer Menge mit Hilfe von Algorithmen beruhen, scharf angegriffen bzw. als widersinnig zurückgewiesen worden (vergleiche etwa ZAREMBA [1]).

In Prüfung der vorliegenden Argumente wird in dieser Arbeit das Konzept der zufälligen Auswahl aus einer endlichen Menge vorgeschlagen und geprüft, wie weit man mit diesen eingeschrankten Monte Carlo-Verfahren

a/ die Schwierigkeiten des Konzepts der zufalligen Auswahl aus einer beliebigen Menge vermeiden kann und

b/ doch wieder alle rechentechnisch relevanten Resultate über klassische Monte Carlo-Verfahren gewinnen kann.

Hierzu werden im ersten Abschnitt der Arbeit die grundlegenden Begriffe eingeführt. Diese werden dann im zweiten Abschnitt in Hinblick auf ihre Bedeutung für die numerische Integration diskutiert. Im dritten Abschnitt werden Resultate über die besprochenen Größen angegeben. Diese Resultate beziehen sich zum Teil auf allgemeine Eigenschaften dieser Größen (Beziehungen untereinander und Zusammenhang zu den n-ten Weiten) und zum Teil auf spezielle Funktionenklassen wie Hölder-Räume, Sobolev-Räume und Funktionen von beschränkter Variation.

1. Einführung der grundlegenden Begriffe und Bezeichnungen

1. Es sei K ein kompakter (Hausdorff-) Raum und C(K) der Raum der stetigen, reellwertigen Funktionen auf K. Die Menge der Radon-Maße auf K bezeichnen wir mit M(K).

Hat $\mu_n \in M(K)$ die Form $\mu_n(f) = \sum_{i=1}^{n} a_i f(x_i)$ (mit $a_i \in R$ und $x_i \in K$), so heißt μ_n eine Quadraturformel mit n Stützstellen (auf K). Die Menge dieser Quadraturformeln bezeichnen wir mit $M_n = M_n(K)$.

Identifizieren wir μ_n mit dem Tupel $(a_1, \ldots, a_n, x_1, \ldots, x_n)$, so gilt $M_n(K) = R^n x K^n$. Versehen mit der Produkttopologie ist $M_n(K)$ ein lokalkompakter Raum. Ist Q_n eine Zufallsvariable mit Werten in $M_n(K)$, so heißt Q_n eine stochastische Quadraturformel mit n Stützstellen (auf K). Die Menge dieser stochastischen Quadraturformeln bezeichnen wir mit $S_n = S_n(K)$.

2. Gegeben sei nun ein $\mu \in M(K)$ und eine Menge B von μ-integrierbaren Funktionen auf K. Wir definieren:

$$e_n(B,\mu) = \inf_{\mu_n \in M_n} \sup_{f \in B} |\mu(f) - \mu_n(f)|,$$

$$\sigma_n(B,\mu) = \inf_{Q_n \in S_n} \sup_{f \in B} (E(Q_n(f) - \mu(f))^2)^{\frac{1}{2}}$$

(hierbei sei E der Erwartungswert einer Zufallsvariablen) und

$$\sigma_n^e(B,\mu) = \inf_{Q_n \in S_n^*} \sup_{f \in B} (E(Q_n(f) - \mu(f))^2)^{\frac{1}{2}}.$$

Hierbei sei S_n^* die Teilmenge derjenigen stochastischen Quadraturformeln aus S_n, die nur endlich viele Werte aus M_n annehmen, wobei jeder Wert eine rationale Wahrscheinlichkeit hat.

(Die betrachteten Größen sind 0 oder positiv, möglicherweise auch $+\infty$. Die Größen $\sigma_n^e(B,\mu)$ werden hier - aus Gründen, die im 2. Abschnitt erklart werden - neu eingeführt.)

3. Wir benützen noch folgende Bezeichnung:

Sind $(a_n)_{n\varepsilon N}$ und $(b_n)_{n\varepsilon N}$ zwei Folgen positiver reeller Zahlen mit $0 < c_1 < \frac{a_n}{b_n} < c_2 < \infty$ für gewisse c_1, c_2 und alle $n \varepsilon N$, so schreiben wir $a_n \asymp b_n$ und sagen:
Die Folgen (a_n) und (b_n) haben dieselbe Konvergenzordnung.

2. Diskussion der grundlegenden Begriffe

1. Die Bedeutung der Größen $e_n(B,\mu)$ liegt auf der Hand: Für gegebenes B, μ und n gibt $e_n(B,\mu)$ die maximale Zahl $|\mu(f) - \sum_{i=1}^{n} a_i f(x_i)|$ an, die als Fehler auftreten kann, wenn man auf der Klasse B das Maß μ ersetzt durch eine möglichst günstige Quadraturformel $\mu_n \varepsilon M_n$.

2. Zur Darstellung der Bedeutung der Größen $\sigma_n(B,\mu)$ für die numerische Integration genügt es, den Fall $K = [0,1]$ und $\mu = \lambda$ zu betrachten. Für die Einheitskugel $\overset{1}{C}([0,1])$ im Raum $C([0,1])$ kann man leicht zeigen:

$$\sigma_n(\overset{1}{C}([0,1]),\lambda) \asymp n^{-\frac{1}{2}}.$$

Als stochastische Quadraturformeln, die diese Konvergenzordnung erreichen, kann man das einfachste Monte Carlo-Verfahren wahlen, das gegeben ist durch:

$Q_n(w) = (a_1(w), \ldots, a_n(w), x_1(w), \ldots, x_n(w))$ mit $a_i(w) = \frac{1}{n}$

für alle i und w und die x_i sind voneinander unabhangig und λ-verteilt auf [0,1].

Dieses und ähnliche Ergebnisse werden nun sehr kontrovers diskutiert - vergleiche dazu die extrem verschiedenen Einschatzungen von COVEOU [1] und ZAREMBA [1]. Wir skizzieren hier kurz die beiden gegensatzlichen Standpunkte:

A) Obiger Satz ist numerisch voll nutzbar:

Man wahle n zufallige Punkte $x_1, \ldots, x_n$ aus [0,1] und berechne $Q_n(w)(f) = \frac{1}{n} \sum_{i=1}^{n} f(x_i)$. Dann ist etwa $3 \cdot (E(Q_n(f) - \lambda(f))^2)^{\frac{1}{2}}$ eine Abschätzung für den Fehler $|Q_n(w)(f) - \lambda(f)|$, den man zu erwarten hat. Zwar kann der wahre Fehler auch größer sein als bei dieser stochastischen Fehlerabschätzung, dies wird aber nur relativ selten der Fall sein.

Die zufällige Auswahl von Punkten geschieht mit Hilfe eines Computers und geeignet ausgewählten Algorithmen.

Diesem Standpunkt scheint folgende Vorstellung zugrundezuliegen: Beliebige W-Maße auf $M_n(K)$ lassen sich mit Hilfe von Computern simulieren, d.h., es gibt Algorithmen, die Elemente von $M_n(K)$ gemäß einer vorgegebenen Verteilung "zufällig" auswählen. "Zufällig" heißt hier zunächst lediglich, daß die Begriffsbildungen der Wahrscheinlichkeitstheorie auf solche Auswahlen anwendbar sein sollen.

B) Von den Kritikern der Monte Carlo-Methode wird folgender Standpunkt entgegengestellt:

Wird eine Zahl (oder eine Quadraturformel) mit Hilfe eines festgelegten Algorithmus bestimmt, so ist eben auch die Zahl bzw. die Quadraturformel festgelegt oder determiniert und damit in keinem vernünftigen Sinne zufällig und stochastischen Untersuchungen zugänglich, da die Wahrscheinlichkeitstheorie nur auf zufällige, nicht aber auf determinierte Ereignisse anwendbar ist.

Wir werden diesen Standpunkt noch diskutieren, zunächst soll aber ein einfaches Argument dafür gegeben werden, daß die unter A) gegebene Vorstellung zumindest in dieser starken Form nicht haltbar ist:

Betrachte wieder $K=[0,1]$, $\mu=\lambda$ und eine beliebige mit dem Computer erzeugte stochastische Quadraturformel $Q_n(w)$. Dann sind jedenfalls die Zahlen $a_i(w)$ und $x_i(w)$ berechenbar, also aus einer abzählbaren Menge. Man kann daher leicht ein quadratisch integrierbares $f:[0,1]\to R$ angeben mit $\lambda(f) = 1$ aber $Q_n(w)(f) = 0$ für alle derartigen Quadraturformeln. Andererseits gilt

$\sigma_n({}^1L_2([0,1],\lambda),\lambda) \asymp n^{-\frac{1}{2}}$, was zeigt, daß die Größen $\sigma_n(B,\mu)$ den rechentechnischen Möglichkeiten nicht angepaßt sind.

3. Die Größen $\sigma_n^e(B,\mu)$ berücksichtigen die soeben angedeutete Kritik, indem hier nur zufällige Stichproben aus endlichen Mengen eingehen.
Allerdings muß noch auf die geschilderte grundsätzlichere Kritik von ZAREMBA [1] u.a. eingegangen werden - diese Kritik besteht ja unabhängig von der Mächtigkeit der Menge, aus der ausgewählt werden soll.
Dieser Kritik scheint folgende Vorstellung zugrundezuliegen:
Es gibt zufällige Ereignisse und determinierte - und dazwischen nichts. Zu den determinierten gehören die Ergebnisse

von Rechnungen. Lediglich auf zufällige Ereignisse läßt sich die Wahrscheinlichkeitstheorie anwenden. Daher ist es unsinnig, auf algorithmisch erzeugte "Zufallszahlen" die Gesetze der Stochastik anzuwenden. Zufallsvariablen lassen sich also prinzipiell nicht auf dem Computer simulieren.

Ich bin der Meinung, daß diese Auffassung eine viel zu starke Einschränkung des Anwendungsgebietes der Stochastik beinhaltet und insbesondere folgende kritischen Punkte offen läßt:

i) Wie unterscheidet man Ereignisse, die zufällig sind, von solchen, die es nicht sind? Zufälligkeit sollte nach dieser Auffassung wohl unabhängig von unserem Wissen eine Eigenschaft der Dinge selbst sein - was immer dies auch heißen mag.

ii) Wie verträgt sich diese Auffassung mit der Tatsache, daß die Stochastik weitgehend angewendet wird auf Bereiche, die nicht als zufällig in einem derart starken Sinn angesehen werden (klassische Physik, Medizin, Sozialwissenschaften)?

Zusammenfassend glaube ich, daß die Größen $\sigma_n^e(B,\mu)$ ein sinnvoller Ersatz sein können für die Größen $\sigma_n(B,\mu)$, die einer - bezogen auf die Anwendungen - zu starken Idealisierung entspringen. Andererseits erscheint es mir unnötig, das Konzept der "stochastischen Fehlerabschätzung" ganz aufgeben zu wollen.

3. Resultate

1. Zunächst sollen Ergebnisse angesprochen werden über Beziehungen zwischen den Größen $e_n(B,\mu)$, $\sigma_n(B,\mu)$ und $\sigma_n^e(B,\mu)$ und über Zusammenhänge mit dem von KOLMOGOROV [1] eingeführten Begriff der n-ten Weite.

Satz 1

Stets gilt $e_n(B,\mu) \geq \sigma_n(B,\mu)$.

Satz 2

a) Es sei B eine symmetrische Menge von beschränkten (meßbaren) Funktionen auf K und $\mu \in M(K)$.

Dann gilt:

$$e_n(B,\mu) \leq 2 \cdot ||\mu|| \cdot d_n(B).$$

Hierbei sei $d_n(B)$ die n-te Weite von B bezüglich $||.||_\infty$, d.h. die Größe $d_n(B) = \inf_{X_n} \sup_{f \in B} \inf_{g \in X_n} ||f-g||_\infty$, wobei X_n alle n-dimensionalen linearen Mannigfaltigkeiten durchlauft.

b) Ist μ positiv und $B \subseteq L_2(K,\mu)$ symmetrisch, so gilt

$$\sigma_{n+1}(B,\mu) \leq d_{n,2}(B,\mu) \cdot ||\mu||^{\frac{1}{2}}.$$

Hierbei bezeichnet $d_{n,2}(B,\mu)$ die n-te Weite von B im Raum $L_2(K,\mu)$.

Satz 3

Es gilt $e_n(B,\mu) \to 0$ genau dann, wenn $\sigma_n^e(B,\mu) \to 0$.

Satz 4

Ist B kompakt bezüglich $||.||_\infty$, und $\mu \in M(K)$, so gilt:

$$\sigma_n^e(B,\mu) = \sigma_n(B,\mu).$$

Bemerkungen zu diesen Satzen:

i) Satz 1 ist trivial.

ii) Satz 2 ist der Arbeit [1] des Verfassers entnommen. Satz 2a) folgt mit Hilfe des Interpolationssatzes, der in SHAPIRO [1] dargestellt wird, Satz 2b) benützt ein Konstruktionsverfahren von stochastischen Quadraturformeln von ERMAKOV, ZOLOTUKHIN [1].

iii) Satz 3 zeigt, daß Monte Carlo-Verfahren (im prazisierten Sinne) nur konvergieren können, wenn auch ein klassisches Verfahren konvergiert.

iv) Satz 4 zeigt, daß die Einschrankung auf endliche Stichproben für den wichtigen Fall einer kompakten Menge B keinerlei Verschlechterung der Konvergenz nach sich zieht.

v) Die obigen Satze lassen sich auf allgemeinere Grundbereiche übertragen. Die Beschrankung auf kompakte Grundbereiche erfolgte, weil sich die folgenden Ergebnisse über konkrete Funktionenklassen auf diesen Fall beziehen.

2. Wir geben nun Ergebnisse an, die sich auf konkrete Funktionenklassen beziehen:

Satz 5

Es sei $W_p^k([0,1]^s)$ die Sobolev-Klasse $\{f:[0,1]^s \to R \mid \sum_{|\alpha|=k} ||D^\alpha f||p \leq 1\}$. Dann gilt:

a) $e_n(W_p^k([0,1]^s),\lambda^s) \asymp \sup_{||\mu||\leq 1} e_n(W_p^k([0,1]^s,\mu) \asymp n^{-\frac{k}{s}}$

für p . k > s.

b) $\sigma_n(W_p^k([0,1]^s),\mu) \asymp n^{-\frac{k}{s}-\frac{1}{2}}$

für $\mu = g \cdot \lambda^s$ mit beschranktem g und $p \geq 2$

c) $\sup_{||\mu||\leq 1} \sigma_n(W_p^k([0,1]^s),\mu) = \sup_{||\mu||\leq 1} \sigma_n^e(W_p^k([0,1]^s),\mu) \asymp$

$\asymp n^{-\frac{k}{s}-\frac{1}{2}}$ für p.k > s und $p \geq 2$.

Satz 6

Ist $C^{k,\alpha}([0,1]^s)$ die Hölder-Klasse

$\{f : [0,1]^s \to \mathbb{R} \mid |D^\beta f(x) - D^\beta f(y)| \le ||x - y||^\alpha$

für alle $\beta = (\beta_1, \ldots, \beta_s)$ mit $|\beta| = k\}$, so gilt:

a) $e_n(C^{k,\alpha}([0,1]^s), \lambda^s) \asymp \sup_{||\mu|| \le 1} e_n(C^{k,\alpha}([0,1]^s), \mu) \asymp n^{-\frac{k+\alpha}{s}}$

b) $\sigma_n(C^{k,\alpha}([0,1]^s), \lambda^s) = \sigma_n^e(C^{k,\alpha}([0,1]^s), \lambda^s) \asymp n^{-\frac{k+\alpha}{s} - \frac{1}{2}}$

c) $\sup_{||\mu|| \le 1} \sigma_n(C^{k,\alpha}([0,1]^s), \mu) = \sup_{||\mu|| \le 1} \sigma_n^e(C^{k,\alpha}([0,1]^s), \mu) \asymp$

$\asymp n^{-\frac{k+\alpha}{s} - \frac{1}{2}}$

Bemerkungen:

i) Die Ergebnisse für e_n im Falle des Lebesgue-Maßes λ^s sind schon bekannt, siehe BAHVALOV[1], SOBOLEV[1], POLOVINKIN [1]. BESOV[1], [2].

Die Ergebnisse für σ_n im Falle des Lebesgue-Maßes sind teilweise bekannt, siehe BAHVALOV[1] und HABER[1].

ii) Für den relativ umfangreichen Beweis dieser Sätze - und auch für eine Übertragung auf Besov-Klassen - verweise ich auf die Arbeit [1] des Verfassers.

3. Es könnte der Eindruck entstehen, daß $\sigma_n(B,\mu)$ stets schneller konvergiert als $e_n(B,\mu)$. Dieser Eindruck wäre falsch, es gilt:

Satz 7

Für die Klasse $B = \{f : [0,1] \to \mathbb{R} \mid \mathrm{Var}(f) \le 1\}$ gilt:

i) $e_n(B,\lambda) = \frac{1}{2n}$

ii) $\frac{1}{2n} \geq \sigma_n(B,\lambda) \geq \frac{1}{\sqrt{32}\, n}$

Bemerkung:

Die Aussage i) wurde von ZUBRZYCKI[1] bewiesen, Aussage ii) in der Arbeit [1] des Verfassers.

4. Es scheint sehr schwierig zu sein, die Konstanten in den Sätzen 5 und 6 gut abzuschätzen oder gar genau auszurechnen. Wir betrachten einen Spezialfall, für den man recht genaue Aussagen machen kann und infolgedessen klassische und stochastische Fehlerabschätzungen gut miteinander vergleichen kann:

Satz 8

Es sei $B_s = \{f: [0,1]^s \to \mathbb{R} \mid |f(x_1, \ldots, x_s) - f(y_1,\ldots,y_s)| \leq \max |x_i - y_i|\}$. Dann gilt:

a) $e_n(B_s,\lambda^s) = \frac{s}{2(s+1)}\, n^{-\frac{1}{s}}$ für $n = m^s$

b) $\lim_n e_n(B_s,\lambda^s) \cdot n^{\frac{1}{s}} = \frac{s}{2(s+1)}$

c) $\sigma_n^e(B_s,\lambda^s) = \sigma_n(B_s,\lambda^s) \leq \frac{1}{\sqrt{12}}\, n^{-\frac{1}{2}-\frac{1}{s}}$ für $n = m^s$.

Bemerkung:

Die Aussagen a) und b) sind bekannt, siehe SUHAREV [1]. Wir skizzieren den Beweis von Satz 8c):

1. Schritt:

Wir betrachten für $n = m^s$ folgende stochastische Quadraturformel:

$Q_n(w) = (a_1(w), \ldots, a_n(w), x_1(w), \ldots, x_n(w))$

mit $a_i = \frac{1}{n}$ für alle i und x_i ist gleichverteilt auf einem Würfel Q_i der Seitenlänge $\frac{1}{m}$, die m^s Würfel bilden eine Partition von $[0,1]^s$.

(Man könnte dieses Verfahren "Verfahren der zufälligen Riemann-Summen" nennen.)

2. Schritt:

Durch einfaches Ausrechnen ergibt sich:

$$(E(Q_n(f) - \lambda^s(f))^2) =$$

$$\frac{1}{m^s} \cdot \sum_{i=1}^{n} \int_{Q_i} \left[f(x) - m^s \int_{Q_i} f(x)\,d\lambda^s\right]^2 d\lambda^s.$$

3. Schritt:

Wichtig ist jetzt der folgende Hilfssatz, den man mit Hilfe von Symmetrisierungstechniken beweisen kann:

Hilfssatz:

Für $f \in B_s$ gilt $\int (f - \int f d\lambda^s)^2 d\lambda^s \leq \frac{1}{12}$,

die Konstante $\frac{1}{12}$ ist bestmöglich.

4. Schritt:

Setzt man die Abschätzung des Hilfssatzes (modifiziert für Funktionen auf den Teilwürfeln) in die Formel vom 2. Schritt ein, so erhält man das gewünschte Ergebnis.

Bemerkung:

Für den Fall s = 3 ergibt sich etwa:

$e_{1000}(B_3, \lambda^3) = 0{,}0375$

$e_{10^9}(B_3, \lambda^3) = 0{,}000375$

und $\sigma_{1000}(B_3, \lambda^3) < 0{,}000913$

$\sigma_{10^6}(B_3, \lambda^3) < 2{,}89\ 10^{-6}$

Dieses Beispiel zeigt sehr deutlich, daß der mittlere Fehler von stochastischen Quadraturverfahren in gewissen Fällen sehr viel kleiner ist als der maximal mögliche von deterministischen Verfahren.

LITERATUR

BAHVALOV, N.S. [1] : Über näherungsweise Berechnung von Integralen (Russisch). Vestnik MGV, Ser. Math. Mech. Astron. Phys. Chem., 4 (1959), 3-18

BESOV, O.V. [1] : Intercellular averages and an error estimate for cubature formulas in Sobolev spaces and their generalizations. Proc. of the Steklov Inst. of Math., 1980. Issue 1, 45-60

BESOV, O. V. [2] : Error estimates for cubature formulas in terms of smoothness of functions.
Proc. of the Steklov Inst. of Math.,1981. Issue 4, 11-23.

COVEYOU, R.R. [1] : Random number generation is too important to be left to chance. Studies in Applied Mathematics, 3 (1969), 70-111 SIAM 1969

ERMAKOW, S. M., ZOLOTUKHIN, V. G. [1] : Polynomial approximations and the Monte Carlo method. Theor. Prob. and Appl., 5 (1960), 428-431

HABER, S. [1] :Sto chastic quadrature formulas. Math. of. comp., 23 (1969), 751-764

KOLMOGOROV, A. N. [1] : Annäherung von Funktionen einer gegebenen Funktionenklasse. Ann. of. Math., 31 (1936), 197-111

NOVAK, E. [1] : Zur unteren Fehlergrenze von Quadraturverfahren Dissertation, Erlangen 1983, xi. + 80 S.

POLOVINKIN, V. I. [1]: Sequences of functionals with a boundary layer. Siberian Math. J., 15 (1974), 296-308

SHAPIRO, H.S. [1]: Topics in approximation theory. Lecture Notes in Math., 187, 1971

SOBOLEV, S. L. [1]: On the order of convergence of cubature formulas. Soviet Math. Dokl., 6 (1965), 808-811

SUHAREV, A.G. [1]: Optimal numerical integration formulas for some classes of functions of several variables. Soviet Math. Dokl., 20 (1979), 472-475

ZAREMBA, S.K. [1]: The mathematical basis of Monte Carlo and Quasi Monte Carlo methods. Studies in Applied Mathematics, 3 (1969), 1-12 SIAM 1969 auch in: SIAM Rev., 10 (1968), 303-314

ZUBRZYCKI, S. [1]: Some approximate integration formulas of statistical interest. Coll. Math., 11 (1963/64), 123-135

Erich Novak
Mathematisches Institut der Universität
Bismarckstraße 1 1/2
D-8520 Erlangen

Proc. of the 4th Pannonian Symp. on Math. Stat., Bad Tatzmannsdorf, Austria 1983
W. Grossmann, G. Pflug, I. Vincze, W. Wertz, eds

MULTIVARIATE ESTIMATION WITH HIGH BREAKDOWN POINT

PETER ROUSSEEUW

Delft

1. INTRODUCTION

Suppose we have a sample $X = \{x_1,\ldots,x_n\}$ of n points in p dimensions, and we want to estimate its location by means of an estimator T which is translation equivariant:

$$T(x_1+b, \ldots, x_n+b) = T(x_1,\ldots,x_n) + b \text{ for all } b \tag{1.1}$$

and permutation invariant:

$$T(x_{\pi(1)},\ldots, x_{\pi(n)}) = T(x_1,\ldots, x_n) \text{ for any permutation } \pi. \tag{1.2}$$

The best known estimator is the arithmetic mean $\bar{x} = \frac{1}{n}\Sigma_{i=1}^n x_i$, which is the least squares estimator because it minimizes $\Sigma_{i=1}^n ||x_i-T||^2$ where $||..||$ is the ordinary Euclidean norm. However, it is well-known that $\bar{x}$ is not robust, because even a single (very bad) outlier in the sample can move $\bar{x}$ arbitrarily far away. To formalize this, we shall make use of the breakdown point ε^* defined by Hampel (1971). In this paper, we use a version introduced by Donoho and Huber (1983) which is intended for finite samples, like the precursor idea of Hodges (1967). Denote by $\beta(m;T,X)$ the supremum of $||T(X')-T(X)||$ for all corrupted samples X′ where m of the original data points are replaced by arbitrary values. Then the breakdown point of the estimator T at the sample X is defined as

$$\varepsilon_n^*(T,X) = \min\{\frac{m}{n};\ \beta(m;T,X) \text{ is infinite}\}. \tag{1.3}$$

In words, it is the smallest fraction of contamination that can cause the estimator to take on values arbitrarily far away from T(X). For the arithmetic mean we find $\varepsilon_n^*(T,X) = \frac{1}{n}$ because one bad observation can already cause breakdown. This ε_n^* depends only slightly on n; in order to have only a single value, one often considers the limit for $n \to \infty$ (with p fixed), so we can say that $\bar{x}$ has a breakdown point $\varepsilon^* = 0$ %.

The maximal breakdown point is 50 % for translation equivariant and permutation invariant estimators (consider a configuration of outliers which is just a translation image of the "good" data points). In one dimension, the median reaches $\varepsilon^* = 50$ %. There are two straightforward generalizations to the case $p > 1$:

(i) the coordinatewise median $(\text{med}_i(x_i^1), \ldots, \text{med}_i(x_i^p))$ where x_i^j is the j-th coordinate of x_i

(ii) the L_1 estimator, which minimizes $\Sigma_{i=1}^n ||x_i - T||$.

Both estimators attain $\varepsilon^* = 50$ %.

<u>Remark 1</u> For $p \geq 3$, the coordinatewise median does not have to lie in the convex hull of the sample. (As an example, consider the p unit vectors $X := \{(1,0,..,0), (0,1,..,0),.., (0,0..,1)\}$, the convex hull of which is a simplex which does not contain the coordinatewise median $(0,0,..,0)$.) However, it does lie in the convex hull for $p \leq 2$, as can be shown by a geometrical argument.

2. AFFINE EQUIVARIANCE

In many situations, one wants the estimation to commute with coordinate transformation. We say that T is affine equivariant if and only if

$$T(Ax_1+b,..,Ax_n+b) = AT(x_1,..,x_n)+b \qquad (2.1)$$

for any vector b and any nonsingular matrix A. The arithmetic mean $\bar{x}$ is affine equivariant, unlike the coordinatewise median and the L_1 estimator. However, the L_1 still satisfies the weaker condition of orthogonal equivariance (this is (2.1) restricted to orthogonal matrices A) because it only depends on distances.

On the other hand, multivariate M-estimators are affine equivariant, but their breakdown point is at most $1/(p+1)$ (see Maronna 1976). Donoho (1982) lists some other well-known affine equivariant estimators, and shows that they all satisfy $\varepsilon^* \leq 1/(p+1)$. This raises the question whether it is at all possible to combine a high breakdown point with affine equivariance. In the next section we shall see that this is indeed the case, by discussing the Stahel-Donoho estimator and a new proposal. However, it seems that the affine equivariance condition is very demanding, because the latter estimators are quite complicated. The remainder of this section will be devoted to illustrating the severeness of this condition.

Remark 2 If T is affine equivariant, permutation invariant and continuous in the observations, then T is the arithmetic mean. This follows from taking A in (2.1) equal to diagonal matrices with (p-1) entries λ and one entry 1 . If one lets λ

tend to zero, it follows by continuity that T has to commute with all projections on coordinates. But then it follows from Donoho (1982, proposition 4.6) that T is the arithmetic mean. (This result can also be proven from section 3 of Obenchain 1971).

Remark 3 Some people have expressed the opinion that it should be very easy to obtain high breakdown estimators which are affine equivariant, by rescaling the observations. They proposed the following procedure: calculate the covariance matrix V, and take a root S (that is, $SS^t = V$). Transform the data as $\tilde{x}_i := S^{-1}x_i$, apply to these $\tilde{x}_i$ an easily computable estimator $\tilde{T}$ with $\varepsilon^* = 50\,\%$ which is not affine equivariant, and then transform back by putting $T := S\tilde{T}$. First, we observe that this construction gives a unique result if and only if $\tilde{T}$ is equivariant for orthogonal transformations, because for any orthogonal matrix 0 the product SO is also a root of V. Also $\tilde{T}$ has to be translation equivariant to insure affine equivariance of T. Therefore, it seems that taking the L_1 estimator for $\tilde{T}$ will do the job. However, we shall see that its good breakdown behaviour does not carry over.

To show this, we consider a two-dimensional example where a fraction $(1-\varepsilon)$ of the data is spherically bivariate normal (we assume that both coordinates follow a univariate normal distribution N(0,1) and are independent of each other), and there is a fraction ε of outliers which are concentrated at the point with coordinates (0,u). Here, $u > 0$ and $0 < \varepsilon < \frac{1}{2}$. Calculation shows that the covariance matrix V is diagonal, with the diagonal elements $(1-\varepsilon)$ and $(1-\varepsilon)(1+\varepsilon u^2)$. Therefore, we can easily construct a root S of V by taking the square root of these diagonal elements, yielding $S_{11} = (1-\varepsilon)^{1/2}$, and $S_{22} = ((1-\varepsilon)(1+\varepsilon u^2))^{1/2}$, and of course $S_{12} = S_{21} = 0$. For u tending to ∞, the transformation $S^{-1} : (x,y) \to (\tilde{x},\tilde{y})$ yields the following situation: the fraction $(1-\varepsilon)$ gets concentrated on the x-axis with univariate normal distribution $N(0,(1-\varepsilon)^{1/2})$, and the fraction ε lands at the point $(0,(\varepsilon(1-\varepsilon))^{-1/2})$. For any finite value of u, the L_1 estimate of the transformed data lies on the vertical axis by symmetry, so we may denote it by $\tilde{T} = (0,\Delta(u))$. Let us now look at the limit $\Delta := \lim_{u\to\infty} \Delta(u)$. By definition, Δ minimizes the expected value of $\|(\tilde{x},\tilde{y}) - (0,\Delta)\|$ which is denoted by $D := \varepsilon((\varepsilon(1-\varepsilon))^{-1/2}-\Delta) + (1-\varepsilon)\int[\Delta^2 + \tilde{x}^2]^{1/2} dF(\tilde{x})$, where F is the normal distribution $N(0,(1-\varepsilon)^{-1/2})$. By the substitution $v = (1-\varepsilon)^{1/2}\tilde{x}$, the average distance D equals $\varepsilon((\varepsilon(1-\varepsilon))^{-1/2}-\Delta) + (1-\varepsilon)^{1/2} \int[(1-\varepsilon)\Delta^2 + v^2]^{1/2} d\Phi(v)$ where Φ is the standard normal distribution. Because Δ minimizes D, it follows that $0 = \partial D/\partial\Delta = -\varepsilon + (1-\varepsilon)\int \{(1-\varepsilon)\Delta^2/[(1-\varepsilon)\Delta^2 + v^2]\}^{1/2} d\Phi(v)$, hence $\int\{(1-\varepsilon)\Delta^2/[(1-\varepsilon)\Delta^2+v^2]\}^{1/2}d\Phi(v)$ equals the positive constant $\varepsilon/(1-\varepsilon)$. This cannot happen for $\Delta = 0$, hence $\Delta > 0$. As the final estimate T equals $S\tilde{T}$, we conclude that $\|T\| = \|S\tilde{T}\| = \|(0,S_{22}(u)\Delta(u))\| = \{(1-\varepsilon)(1+\varepsilon u^2)\}^{1/2}\Delta(u)$ tends to infinity for $u \to \infty$. (Note that this effect is due to the explosion of the first factor, caused by the breakdown of S !) This means that any fraction $\varepsilon > 0$ can make T break down, so $\varepsilon^*(T) = 0$ %. Therefore, rescaling does not solve our problem unless we could start with a high-breakdown covariance estimator, but that is precisely what we are looking for in the first place.

3. HIGH-BREAKDOWN ESTIMATORS

The first affine equivariant multivariate location estimator with a breakdown point of 50 % was obtained independently by W. Stahel (1981) and D. Donoho (1982). This estimator, called "outlyingness-weighted mean", is defined as follows. For each observation x_i in X, one looks for a one-dimensional projection in which x_i is most outlying:

$$r_i := \sup_{||u||=1} \frac{u^t x_i - \operatorname{med}_j(u^t x_j)}{\operatorname{med}_k |u^t x_k - \operatorname{med}_j(u^t x_j)|} \tag{3.1}$$

where $\operatorname{med}_j(u^t x_j)$ is the median of all projections of data points x_j on the direction of the vector u. Then one estimates location by the weighted mean

$$T(x_1,\ldots,x_n) := \frac{\Sigma_{i=1}^n \, w(r_i)x_i}{\Sigma_{i=1}^n \, w(r_i)} \tag{3.2}$$

where w(r) is a strictly positive and decreasing function of $r \geq 0$, such that $r \cdot w(r)$ is bounded. (The bound on $r \cdot w(r)$ is motivated by the one-dimensional case, where (3.2) is a one-step W-estimator starting from the median (Hoaglin et al 1983, page 291), the influence function of which is proportional to $r \cdot w(r)$.) Formula (3.1) is related to projection pursuit (Friedman and Tukey 1974), because one really has to search over all possible directions to find the "best" projection. It is shown that (3.2) is affine equivariant, and has a breakdown point converging to 50 % (Donoho 1982, proposition 3.1) at samples X in <u>general position</u>, which means that no more than p points of X lie in any (p-1) dimensional affine subspace. (3.3)

For two-dimensional data (visualized as points in a plane) this means that there are no more than two points of X on any line, so any three points of X determine a triangle with nonzero area.

Let us now introduce a second affine equivariant estimator with high breakdown, by putting

$$T(X) := \text{center of the minimal volume ellipsoid covering (at least) h points of X} \quad (3.4)$$

where $h = [\frac{n}{2}]+1$ (here $[\frac{n}{2}]$ is the largest integer $\leq \frac{n}{2}$). We call this the Minimal Volume Ellipsoid estimator (MVE). The affine equivariance of the MVE follows from the fact that for each ellipsoid E the image f(E) through the nonsingular affine transformation $f(x) = Ax+b$ is again an ellipsoid, and

$$\text{Volume } (f(E)) = |\det (A)| \text{ Volume } (E).$$

Because det(A) is a constant, the relative sizes of ellipsoids do not change under affine transformations.

<u>Proposition 3.1</u> At any p-dimensional sample X in general position, the breakdown point of the MVE estimator equals

$$\varepsilon_n^*(T,X) = ([\tfrac{n}{2}]-p+1)/n$$

which converges to 50 % as $n \to \infty$.

Proof Without loss of generality, let $T(X) = 0$. We put $M :=$ volume of smallest ellipsoid with center zero containing all points of X. Because X is in general position, each of its $\binom{n}{p+1}$ subsets S_j of $p+1$ points determines a simplex with nonzero volume. Therefore, for each S_j there exists a bound d_j such that any ellipsoid with center $\|c\| > d_j$ and containing S_j has a volume strictly larger than M. Put $d := \max_j d_j < \infty$.

Let us first show that $\varepsilon_n^*(T,X) \leq ([\frac{n}{2}]-p+1)/n$. Take any sample X' obtained by replacing at most $[\frac{n}{2}]-p$ points of X. Suppose $\|T(X')\| > d$, and let E be the corresponding smallest ellipsoid containing (at least) $[\frac{n}{2}]+1$ points of X'. But then E contains at least $([\frac{n}{2}]+1) - ([\frac{n}{2}]-p) = p+1$ points of X, so Volume $(E) > M$. This is a contradiction, because the smallest ellipsoid with center zero around the $n-([\frac{n}{2}]-p) \geq [\frac{n}{2}]+1$ "good" points of X' has a volume of at most M. Therefore, $\|T(X')\| \leq d$. (Even if $T(X')$ is not unique, then $\|T(X')\| \leq d$ still holds for all solutions.)

On the other hand, $\varepsilon_n^*(T,X) \leq ([\frac{n}{2}]-p+1)/n$. Indeed, take any p points of X, and consider the $(p-1)$-dimensional affine subspace H they determine. Now replace $[\frac{n}{2}]-p+1$ other points of X by points on H. Then H contains $[\frac{n}{2}]+1$ points of the new sample X', so the minimal volume ellipsoid covering these $[\frac{n}{2}]+1$ points degenerates to zero volume. Because X is in general position, no other ellipsoid covering $[\frac{n}{2}]+1$ points of X' can have zero volume, so $T(X')$ lies on H. Finally, we note that $T(X')$ is not bounded because the $[\frac{n}{2}]-p+1$ contaminated data points on H may have arbitrarily large norm. This ends the proof. □

A variant of the MVE is the estimator

$$T(X) := \text{mean of the h points of X for which the determinant of the covariance matrix is minimal.} \quad (3.5)$$

We call this the Minimal Covariance Determinant estimator (MCD), which corresponds to finding the h points for which the confidence ellipsoid (for any given level) has minimal volume, and then taking its center. This estimator is also affine equivariant because the determinant of the covariance matrix of the transformed data equals

$$\det(AVA^t) = \det^2(A) \det(V).$$

The MCD also has the same breakdown point as the MVE, using the same reasoning as in Proposition 3.1.

Both the MVE and the MCD are very drastic, because they are intended to safeguard against up to 50 % of outliers. If one is certain that the fraction of outliers is at most α (where $0 < \alpha \leq \frac{1}{2}$), one can work with the estimators MVE(α) and MCD(α) obtained by replacing h by $k(\alpha) := [n(1-\alpha)] + 1$ in (3.4) and (3.5). The breakdown point of these estimators is equal to α (for $n \to \infty$). For $\alpha \to 0$, the MVE yields the center of the smallest ellipsoid covering all the data, whereas the MCD tends to the arithmetic mean.

Also note that both the MVE(α) and the MCD(α) yield very robust covariance estimators at the same time, if one makes use of the selected fraction $k(\alpha)$ of observations and multiplies the resulting matrix with a constant to obtain consistency in the case of multivariate normality.

4. A PARTICULAR CASE

In the general multivariate setting, the asymptotic properties of the MVE and MCD estimators are hard to determine. However, a beginning can be made by considering the special case where p=1. In this situation, it is easy to give an explicit algorithm for calculating the MVE and the MCD estimates. First one orders the data, so we may assume that

$$x_1 \le x_2 \le \ldots \le x_n.$$

Only n-h+1 "halves" of this sample need to be considered, namely

$$H_1 = \{x_1, \ldots, x_h\},\ H_2 = \{x_2, \ldots, x_{h+1}\} \ \ldots$$

$$\ldots,\ H_{n-h+1} = \{x_{n-h+1}, \ldots, x_n\}.$$

For each of these H_i, one defines

$$\text{midpoint}(i) = \tfrac{1}{2}(x_i + x_{i+h-1}) \quad \text{length}(i) = x_{i+h-1} - x_i$$

$$\text{mean}(i) = \frac{1}{h} \Sigma_{j=i}^{i+h-1}\, x_j \quad \text{var}(i) = \frac{1}{h-1} \Sigma_{j=i}^{i+h-1} (x_j - \text{mean}(i))^2.$$

For the MVE one selects the H_i for which length(i) is minimal, and one puts T(X):=midpoint(i). (In case more than one length(i) is minimal, T(X) can be taken as the average of the corresponding numbers midpoint(i).) For the MCD, one selects the H_i for which var(i) is minimal and puts T(X):=mean(i); if there are several solutions, one again can take their average. Also the MVE(α) and MCD(α) estimates can be calculated this way, by replacing h by $k(\alpha) = [n(1-\alpha)]+1$.

Remark 4 When p=1, the MVE(α) estimator can be rewritten as

$$\underset{t}{\text{minimize}}\ (r^2)_{k(\alpha):n} \tag{4.1}$$

where the $(r^2)_{i:n}$ are the squared residuals $(x_j-t)^2$, ordered from smallest to largest. The MCD(α) estimator is the solution of

$$\underset{t}{\text{minimize}}\ \Sigma_{i=1}^{k(\alpha)} \quad (r^2)_{i:n}.$$

Both criteria (4.1) and (4.2) can also be applied to the estimation of regression coefficients, and yield a breakdown point equal to α in that case too (Rousseeuw 1982, 1983).

Let us now consider a symmetric and strongly unimodal distribution F with density f and finite Fisher information.

Proposition 4.1 If the observations are iid according to $F(x-\theta)$, then the distribution of the MVE(α) estimator T_n converges as:

$$\mathcal{L}(n^{1/3}(T_n-\theta)) \xrightarrow{\text{weak}} \mathcal{L}(c\tau/f(F^{-1}(1-\tfrac{\alpha}{2})).$$

Here $c=(\frac{1}{2}\Lambda^2(F^{-1}(1-\frac{\alpha}{2})))^{-1/3}$ where $\Lambda=-f'/f$ corresponds to the maximum likelihood scores, and τ is the random time s for which $s^2+Z(s)$ attains its minimum, where Z(s) is a standard Brownian motion.

Proof The proof is a simple adaptation of the result for a very similar estimator, namely the "shorth" (=mean of the shortest half of the sample). Parts 1, 2 and 3 of the reasoning in (Ahdrews et al. 1972, page 51) still apply and yield the constant c. We slightly adapt the remaining part of the calculation, using the same notations. If $\theta=0$ and $\hat{t}$ is the minimizing value of t, then main asymptotic variability of T_n is given by

$\frac{1}{2}\left[F_n^{-1}((\frac{1}{2}+\hat{t})+(\frac{1-\alpha}{2}))+F_n^{-1}((\frac{1}{2}+\hat{t})-(\frac{1-\alpha}{2}))\right] \cong \hat{t}(F^{-1})'(\frac{1}{2}+(\frac{1-\alpha}{2})) =$ $\hat{t}/f(F^{-1}(1-\frac{\alpha}{2}))$, where $n^{1/3}\hat{t}$ behaves asymptotically like $c\tau$. □

This slow rate of convergence is a disadvantage of the MVE(α) method. However, note that number of arithmetic operations needed to compute MVE(α) is of the order $O(n\log(n))$, which is much faster than the $O(n^2)$ needed for the MCD(α) estimator.

Proposition 4.2 If the observations are iid according to $F(x-\theta)$, then the MCD(α) estimator T_n is asymptotically normal:

$$\mathcal{L}\,(n^{1/2}(T_n-\theta)) \xrightarrow{\text{weak}} N(0,V(\alpha))$$

where the asymptotic variance aquals

$$V(\alpha) = \frac{2\int_0^q x^2\, dF(x)}{(1-\alpha-2qf(q))^2}$$

(here, q is the quantile $F^{-1}(1-\frac{\alpha}{2})$).

Proof For each t, denote by $R_i(t)$ the rank of $|x_i-t|$. Moreover, the function a on $[0,1]$ is defined by $a(u)=(1-\alpha)^{-1}$ if $u \leq 1-\alpha$, and $a(u)=0$ otherwise. Then the MCD(α) estimator corresponds to $\underset{t}{\text{minimize}}\ \Sigma_{i=1}^{n}\ a(R_i(T)/n)(x_i-t)^2$. As there exists only a finite number of points t where some $R_i(t)$ changes, we may assume w.l.o.g. that all $R_i(t)$ are locally constant around the solution $t = T_n$. On this neighbourhood $(T_n-\delta,T_n+\delta)$ we put $a_i:=(R_i(T_n)/n)$ and

$S_n := \frac{1}{n}\Sigma_{i=1}^n a(R_i(T_n)/n)x_i$. Now T_n minimizes

$$\Sigma_{i=1}^n a_i((x_i-S_n)+(S_n-t))^2 = \Sigma_{i=1}^n a_i(x_i-S_n)^2 + \Sigma_{i=1}^n a_i(S_n-t)^2$$

for all t in

$(T_n-\delta, T_n+\delta)$. Differentiating this expression with respect to t at T_n, we find that $T_n=S_n$. Therefore, T_n is a solution of the equation

$$\frac{1}{n}\Sigma_{i=1}^n a(R_i(t)/n)x_i = t. \tag{4.3}$$

Estimators satisfying (4.3) with this function a go back to Gnanadesikan and Kettenring (1972), and their asymptotic normality was shown by Yohai and Maronna (1976), giving the desired result. □

<u>Remark 5</u> This proposition implies that MCD(α) has the same asymtotic behaviour at F as the Huber-type "skipped" mean (Hampel 1974, page 392), which is the M-estimator of location (Huber 1981) corresponding to the ψ function

$$\begin{aligned} \psi(x) &= x \quad && |x| \leq F^{-1}(1-\tfrac{\alpha}{2}) \\ &= 0 && \text{elsewhere} \end{aligned} \tag{4.4}$$

<u>Remark 6</u> The results of this section imply that our 50 % breakdown estimators MVE and MCD possess low asymptotic efficiencies. However, this can be repaired by applying a one-step M-estimator afterwards (Huber 1981, page 146), preferably using a smooth redescending ψ-function such as the one proposed by Hampel, Rousseeuw and Ronchetti (1981). In this way, the asymptotic efficiency can be improved without losing $\varepsilon^* = 50$ %.

Acknowledgments This research was supported by the Belgian National Science Foundation, and most of it took place while the author was visiting the Mathematical Sciences Research Institute at Berkeley. Thanks are due to D. Donoho for valuable suggestions and stimulating discussions. Helpful comments were made by F. Hampel, A. Leroy, F. Plastria, W. Stahel and V. Yohai.

REFERENCES

ANDREWS, D., BICKEL, P., HAMPEL, F., HUBER, P., ROGERS, W., and TUKEY, J. (1972).: Robust Estimates of Location: Survey and Advances. Princeton University Press.

DONOHO, D. (1982)."Breakdown Properties of Multivariate Location Estimators", Ph.D. qualifying paper, Harvard University.

DONOHO, D. and HUBER, P. (1983). "The Notion of Breakdown Point", in: A Festschrift for Erich L. Lehmann, Wadsworth.

FRIEDMAN, J. and TUKEY, J. (1974). "A Projection Pursuit Algorithm for Exploratory Data Analysis", IEEE Trans. Comp. C-23, 881-889.

GNANADESIKAN, R. and KETTENRING, J. (1972). "Robust Estimates, Residuals, and Outlier Detection with Multiresponse Data" Biometrics 28, 81-124.

HAMPEL, F. (1971). "A General Qualitative Definition of Robustness", Ann. Math. Statist. 42, 1887-1896.

HAMPEL, F. (1974). "The Influence Curve and its Role in Robust Estimation", JASA 69, 383-393.

HAMPEL, F. ROUSSEEUW, P., and RONCHETTI, E. (1981). "The Change-of-Variance Curve and Optimal Redescending M-estimators," JASA 76, 643-648.

HOAGLIN, D.,MOSTELLER, F., TUKEY, J. (1983). Understanding Robust and Exploratory Data Analysis. Wiley, New York.

HODGES, J. (1967). "Efficiency in Normal Samples and Tolerance of Extreme Values for some Estimates of Location", Proc. 5th Berkeley Symp. on Math. Stat. and Prob., Vol. 1, 163-168.

HUBER, P. (1981). Robust Statistics. Wiley, New York.

MARONNA, R. (1967). "Robust M-estimators of Multivariate Location and Scatter", Ann. Statist. 4. 51-67.

OBENCHAIN, R. (1971). "Multivariate Procedures Invariant under Linear Transformations", Ann. Math. Statist. 42. 1569-1578.

ROUSSEEUW, P. (1982). "Least Median of Squares Regression", Technical Report, University of Brussels (to appear in JASA).

ROUSSEEUW, P. (1983). "Regression Techniques with High Breakdown Point", IMS Bulletin 12, page 155.

STAHEL, W. (1981). "Breakdown of Covariance Estimators" Research Report No. 31, Fachgruppe für Statistik, ETH Zürich.

YOHAI, V. and Maronna, R. (1976). "Location Estimators based on Linear Combinations of Modified Order Statistics", Commun. Statist. A5, 481-486.

Peter Rousseeuw
Delft University of Technology
Dept. of Mathematics and Informatics
Julianalaan 132
2628 BL Delft
The Netherlands

Proc. of the 4th Pannonian Symp. on Math. Stat., Bad Tatzmannsdorf, Austria 1983
W. Grossmann, G. Pflug, I. Vincze, W. Wertz, eds

ROBUST ESTIMATION OF AUTOREGRESSIVE MOVING AVERAGE MODELS

NORBERT STOCKINGER

Graz

1. Introduction

Consider the class of time series given by an *autoregressive moving average model* of order p and q (short ARMA(p,q) model)

$$y_i = f_i(\underline{\alpha}) + a_i, \quad i = \ldots,0,1,2,\ldots \tag{1}$$

where $\underline{\alpha}^T = (\alpha_1,\ldots,\alpha_{1+p+q}) = (\lambda, \underline{\phi}^T, \underline{\theta}^T)$

and $f_i(\underline{\alpha}) = \lambda + (y_{i-1},\ldots,y_{i-p})\underline{\phi} - (a_{i-1},\ldots,a_{i-q})\underline{\theta}$

and where a_i, the so-called innovations, are i.i.d. random variables with scale σ_a and expectation 0. λ denotes the intercept, $\underline{\phi}^T = (\phi_1,\ldots,\phi_p)$ the vector of autoregressive parameters and $\underline{\theta}^T = (\theta_1,\ldots,\theta_q)$ the vector of moving average parameters.

The model (1) is called *outlier free* if the a_i are normally distributed and it is called *innovations outlier model* if the a_i have a heavy-tailed distribution.

Suppose that we have observed the variables

$$y_i \text{ for } i = 1,\ldots,n, \tag{2}$$

and the parameter vector $\underline{\alpha}$ is to be estimated. We see immediately that the $f_i(\underline{\alpha})$ depend on the unknown realizations of the previous innovations a_{i-1}, a_{i-2},... which in turn might be replaced by the residuals $r_{i-1} = y_{i-1} - f_{i-1}(\underline{\alpha}),\ldots$.

The ARMA parameters $\underline{\alpha}$ in (1) are conventionally, e.g. in the BOX and JENKINS Time Series Package, estimated by the method of least-squares. Note that the estimation problem is nonlinear (i.e. the residuals depend nonlinearly on the parameters) for $q > 0$, because the f_i depend on residuals which themselves depend on the parameter vector $\underline{\alpha}$. Properties, such as consistency, asymptotic normality and robustness of least-squares estimates are mainly investigated for autoregressive models, i.e. $q = 0$ in (1), and can e.g. be found in MARTIN (1979, 1980) and YOHAI and MARONNA (1977). In any case the theoretical optimum properties are only assured under the assumption of a Gaussian distribution of the innovations.

The idea to get a reliable estimate of $\underline{\alpha}$ if the innovations are heavy-tailed distributed is to bound the influence of the residuals in the minimum problem defining an estimate of $\underline{\alpha}$, what is performed in a *maximum-likelihood type estimate* (short M--estimate) $\hat{\underline{\alpha}}_M$. This is defined to be a solution of

$$g(\underline{\alpha}', s) = \sum_{i=p+1}^{n} \rho\left(\frac{y_i - f_i(\underline{\alpha}')}{s}\right) = \min \tag{3}$$

where s is a robust estimate of the innovations scale σ_a and ρ is a symmetric robustifying loss function (HUBER, 1973). ρ is often given in the form of its derivative ψ. ψ-functions used in this paper are *HUBER's monotone psi-function*

$$\psi_H(t) = \operatorname{sgn}(t) \min(|t|, c) \tag{4}$$

where c is a contant and

HAMPEL's three part redescending psi-function

$$\psi_{HA}(t) = \begin{cases} t & ,|t| \le a \\ a \text{ sgn } (t) & , a < |t| \le b \\ a(t-d \text{ sgn}(t))/(b-d) & , b < |t| \le d \\ 0 & , d < |t|. \end{cases} \tag{5}$$

Other ψ-functions are collected in DUTTER (1980). MARTIN (1981) mentions the possibility of M-estimation for ARMA models, but does not give a computational procedure which is the main objective of the presented paper.

Following a proposal of HUBER (1973) σ_a and $\underline{\alpha}$ may be found simultaneously through solving (3) and the side condition

$$\frac{1}{n-2p-q-1} \sum_{i=p+1}^{n} \psi^2\left(\frac{y_i - f_i(\underline{\alpha}')}{s}\right) = B \tag{6}$$

if a monotone psi-function - like ψ_H - is used. The constant B is selected so that s is asymptotically unbiased for σ_a when the innovations distribution is Gaussian with mean zero and standard deviation σ_a, i.e., $B = E_\Phi \psi^2(r)$ where Φ denotes the standard normal distribution and r is a random variable with distribution Φ.

The minimum problem (3) and the side condition (6) can be combined to a more general minimum problem

$$e(\underline{\alpha}', s) = \sum_{i=p+1}^{n} \rho\left(\frac{y_i - f_i(\underline{\alpha}')}{s}\right)s + As = \min \tag{7}$$

where the minimum in both $\underline{\alpha}'$ and s is to be achieved and

$A = (n-2p-q-1)B/2$. This minimum problem can be solved iteratively by the WS-algorithm (W-Sophisticated) given by DUTTER and HUBER (1981). The WS-algorithm is formed by the iterated weighted least-squares (W-)algorithm, which can e.g. be found in STOCKINGER and DUTTER (1983), and the algorithm of NAGEL and WOLFF (1974) for solving the nonlinear least-squares problem. The algorithm of NAGEL and WOLF is based on a linar compromise between the Gauss-Newton procedure and the method of steepest descent. In order to estimate ARMA parameters $\underline{\alpha}$ the WS-algorithm can be used in the following way.

2. The WS-algorithm

We have to make the following restrictions: the function ρ is an even function (i.e. $\rho(t) = \rho(-t)$ for all $t \in \mathbb{R}$), ρ'' is bounded, $0 \leq \rho'' \leq \rho''_0$ $\psi(t)/t$ is monotone nonincreasing for $t>0$ and $\lim_{t\to 0} \psi(t)/t = \rho''_0 < \infty$. We need starting values $\underline{\alpha}^{(0)}$ and $s^{(0)}$ a tolerance level $\varepsilon > 0$ and the constant A as defined in (7). If $q > 0$ we set $r^{(k)}_{p+1-q} = \ldots = r^{(k)}_p = 0$ for all $k = 0,1,\ldots$.

The residuals $\underline{r}^{(k)} = (r^{(k)}_{p+1}, \ldots, r_n^{(k)})^T$ which have to be computed for the WS-algorithm are here computed in the following successive manner.

(1) Put $i = p$.

(2) Augment $i = i+1$.

(3) Compute $r_i^{(k)} = y_i - (1,y_{i-1},\ldots,y_{i-p},-r^{(k)}_{i-1},\ldots,-r^{(k)}_{i-q})\underline{\alpha}^{(k)}$.

(4) If $i < n$ go to (2).

Note that the value of the function g given in (3) (an estimate of $\underline{\alpha}$ has to minimize g) at $\underline{\alpha}^{(k)}$ and $s^{(j)}$ for some k and j is computed by

$$\sum_{i=p+1}^{n} \rho(r_i^{(k)}/s^{(j)}).$$

We propose the algorithm following DUTTER and HUBER (1981) where the motivation of the steps may be seen.

1. Put $k = 0$.
2. Compute residuals $\underline{r}^{(0)}$.
3. Find an improved scale

$$(s^{(k+1)})^2 = \frac{1}{A} \sum_{i=p+1}^{n} \chi\left(\frac{r_i^{(k)}}{s^{(k)}}\right) (s^{(k)})^2$$

   where $\chi(t) = t\psi(t) - \rho(t)$.

4. Calculate the weights

$$w_i^{(k)} = \psi(r_i^{(k)}/s^{(k+1)})/(r_i^{(k)}/s^{(k+1)}),$$

   if $r_i^{(k)} \neq 0$, otherwise $w_i^{(k)} = 1$, $i = p+1,\ldots,n$; define a diagonal matrix $W^{(k)}$ with $w_i^{(k)}$ as its $(i-p)$-th diagonal element.

5. Find $c_{i,j}^{(k)} = \dfrac{\partial f_i(\underline{\alpha}^{(k)})}{\partial \alpha_j}$, $j = 1,\ldots,1+p+q$, $i = p+1,\ldots,n$, successively according to the algorithm given below. Denote the matrix $(c_{i,j}^{(k)})$ by $C^{(k)}$ and $(c_{i,1}^{(k)},\ldots,c_{i,1+p+q}^{(k)})^T$ by $\underline{c}_i^{(k)}$. (Note that from $r_{p+1-q}^{(k)} = \ldots = r_p^{(k)} = 0$ it follows that $c_{i,j}^{(k)} = 0$, $j = 1,\ldots,1+p+q$, $i = p+1-q,\ldots,p$.)

   (1) Put $i = p$.

   (2) Augment $i = i+1$.

   (3) Compute

$$c_{i,1}^{(k)} = 1 + \theta_1^{(k)} c_{i-1,1}^{(k)} + \ldots + \theta_{q,}^{(k)} c_{i-q,1}^{k},$$

$c_{i,j}^{(k)} = y_{i+1-j} + \theta_1^{(k)} c_{i-1,j}^{(k)} + \ldots + \theta_q^{(k)} c_{i-q,j}^{(k)}$, for $2 \le j \le p+1$,

$c_{i,j}^{(k)} = - r_{i+1+p-j}^{(k)} + \theta_1^{(k)} c_{i-1,j}^{(k)} + \ldots + \theta_q^{(k)} c_{i-q,j}^{(k)}$, for $p+2 \le j \le 1+p+q$.

(4) If $i < n$ go to (2).

6. Solve

$$\sum_{i=p+1}^{n} (r_i^{(k)} - \underline{c}_{i.}^{(k)T} t^{(k)})^2 \, w_i^{(k)} = \text{minimum}$$

for the (approximative) direction $\underline{t}^{(k)}$ of the Gauss-Newton method, i.e. solve

$$H^{(k)} \, \underline{t}^{(k)} = \underline{\gamma}^{(k)}$$

where $H^{(k)} = C^{(k)T} \, W^{(k)} \, C^{(k)}/s^{(k+1)}$

and $\underline{\gamma}^{(k)} = C^{(k)T} \, W^{(k)} \underline{r}^{(k)}/s^{(k+1)}$ is the vector of the steepest descent.

7. Calculate $g(\underline{\alpha}^{(k)}, s^{(k+1)})$.

Compute new values for $\underline{\alpha}$ by $\underline{\alpha}^{(k+1)} = \underline{\alpha}^{(k)} + \underline{t}^{(k)}$.

Calculate $\underline{r}^{(k+1)}$ and $g(\underline{\alpha}^{(k+1)}, s^{(k+1)})$.

If $g(\underline{\alpha}^{(k+1)}, s^{(k+1)}) < g(\underline{\alpha}^{(k)}, s^{(k+1)})$ go to 8.

7.a. Compute $\omega = \dfrac{\underline{\gamma}^{(k)T} \, \underline{\gamma}^{(k)}}{\underline{\gamma}^{(k)T} H^{(k)} \underline{\gamma}^{(k)}}$, which is an approximative value

such that g takes its minimum in the direction of

$\mu\omega\underline{\gamma}^{(k)} + (1-\mu)\underline{t}^{(k)}$ with $0 < \mu \le 1$.

Perform the following steps.

(1) Put $\nu = 0$.

(2) Augment $\nu = \nu+1$.

(3) Put $\underline{\alpha}^{(k+1)} = \underline{\alpha}^{(k)} + \frac{\nu}{5} \omega\underline{\gamma}^{(k)} + (1 - \frac{\nu}{5})t^{(k)}$.

(4) Calculate $\underline{r}^{(k+1)}$ and $g_\nu = g(\underline{\alpha}^{(k+1)}, s^{(k+1)})$.

(5) If $g_\nu < g(\underline{\alpha}^{(k)}, s^{(k+1)})$ go to 7b.

(6) If $\nu < 5$ go to (2), else go to 7c.

7.b. Compute a linear back-interpolation between $(\nu - 1)/5$ and $\nu/5$ by

$$\mu^* = \frac{\nu}{5} - \frac{g(\underline{\alpha}^{(k)}, s^{(k+1)}) - g_\nu}{g_{\nu-1} - g_\nu} \frac{1}{5} .$$

Put $\underline{\alpha}^{(k+1)} = \underline{\alpha}^{(k)} + \mu^* \omega \underline{\gamma}^{(k)} + (1-\mu^*)\underline{t}^{(k)}$.

Calculate residuals $\underline{r}^{(k+1)}$ and $g(\underline{\alpha}^{(k+1)}, s^{(k+1)})$.

If $g(\underline{\alpha}^{(k+1)}, s^{(k+1)}) < g(\underline{\alpha}^{(k)}, s^{(k+1)})$ go to 8.

Otherwise set $\underline{\alpha}^{(k+1)} = \underline{\alpha}^{(k)} + \frac{\nu}{5} \omega \underline{\gamma}^{(k)} + (1- \frac{\nu}{5} \underline{t})^{(k)}$.

Compute residuals $\underline{r}^{(k+1)}$ and go to 8.

7.c. Put $\omega = \omega/2$ and compute

$$\underline{\alpha}^{(k+1)} = \underline{\alpha}^{(k)} + \omega \underline{\gamma}^{(k)}.$$

Calculate residuals $\underline{r}^{(k+1)}$ and $g(\underline{\alpha}^{(k+1)}, s^{(k+1)})$.

If $g(\underline{\alpha}^{(k+1)}, s^{(k+1)}) \geq g(\underline{\alpha}^{(k)}, s^{(k+1)})$ start again at 7c.

8. Stop, if

$$|s^{(k+1)} - s^{(k)}| < \varepsilon\, s^{(k+1)}$$

and if the difference between the parameters is less than ε times their approximate standard deviation, i.e., if

$$|\alpha_i^{(k+1)} - \alpha_i^{(k)}| < \varepsilon\, s^{(k+1)} \sqrt{c^{ii}} \quad \text{for all } i = 1,\ldots,1+p+q,$$

where c^{ii} is the i-th diagonal element of $(s^{(k+1)} H^{(k)})^{-1}$.

9. Augment $k = k+1$, go to 3.

If a redescending ψ-function is used, $e(\underline{\alpha}', s)$ in (7) has only to be minimized with respect to $\underline{\alpha}'$ and s must be constant. This minimization can be achieved by modifying the algorithm given above in the following way: the scale $s^{(k)}$ which depends on the iteration counter k has to be replaced by a constant s and the step 3 for finding an improved scale and the convergence criterion for the scale in step 6 has to be eliminated.

3. Illustration

The presented algorithm to estimate ARMA parameters robustly has been approved in a small preliminary study from which the following example will illustrate the behavior of least--squares estimates and M-estimates. First an ARMA(1,1) model with 212 observations, normally distributed innovations and parameters $\sigma_a = .5$, $\lambda = 0$, $\phi = \phi_1 = .25$ and $\theta = \theta_1 = -.75$ is simulated. The least-squares estimates $s = .545$, $\hat{\lambda} = .094$, $\hat{\phi} = .267$ and $\theta = -.748$ for the parameters of that simulated model are a good approximation of the true values, because the model is outlier free. Next the ARMA(1,1) model is contaminated by large innovations embedded at positions i = 50, 90 and 160 (size -2.5, 3.5, -3., respectively), what results in a kind of innovations outlier model. Least-squares estimates for the innovations outlier model are shown in the table below in the row labelled LS. The bad influence of the large innovations can be seen from the fact that the least-squares estimates of the contaminated model are a worse approximation of the true values than least-squares estimates of the outlier free model. The starting values of each least-squares fit were 1 for σ_a, 0 for λ and 0.1 for ϕ and θ. A robust estimation using ψ_H (with c = 1.5

and the least-squares estimates as starting values) is less affected by the large innovations and yields results - presented in the following table -, which are closer to the true values than least-squares estimates. An estimation with ψ_{HA} (with $a = 1.5$, $b = 2.5$ and $d = 5.$), where the results of the previous estimation serve as starting values leads to further improved estimated parameter values, which are also listed in the following table. The three rightmost columns of the following table contain the residuals (computed from the estimated parameters) at positions $i = 50$, 90 and 160. It can be seen, that the absolute values of the residuals computed from robustly estimated parameters are larger than the absolute values of the residuals obtained by a least-squares estimation; this again shows that the robust estimates are less influenced by outliers.

Kind of Estimation	σ_a	λ	ϕ	θ	Residuals at position		
True Values	.5	0.	.25	-.75	50	90	160
LS	.653	.078	.188	-.813	-2.538	3.355	-3.006
ψ_H	.566	.087	.219	-.785	-2.544	3.399	-3.031
ψ_{HA}	.566	.092	.230	-.771	-2.551	3.418	-3.041

It is noted that the presented results are only preliminary and many details, e.g. theoretical properties, need further investigation. This is the subject of present research.

REFERENCES

BOX G.E.P. and G.M. JENKINS: Computer Programs for the Analysis of Univariate Times Series Using the Methods of BOX and JENKINS. Supplementary Program Series No. 517. The University of Wisconsin Computer Center.

DUTTER R. (1980): Robuste Regression: Bericht No. 135, Mathematisch Statistische Sektion im Forschungszentrum Graz.

DUTTER R. and P.J. HUBER (1981): Numerical Methods for the Nonlinear Robust Regression Problem. J. Statist. Comput. Simul. 13, 2, 79-114.

HUBER P. J. (1973): Robust Regression: Asimptotics, Conjectures and Monte Carlo. Ann. Statist., 1, 5, 799-821.

MARTIN R.D. (1979): Robust Estimation for Time Series Autoregressions. In: LAUNER R.L. and G.N. WILKINSON (Eds.), Robustness in Statistics. Academic Press. New York.

MARTIN R.D. (1980): Robust Estimation of Autoregressive Models In: BRILLINGER and TIAO (Eds.), Directions in Time Series. Inst. Math. Statist. Publication.

MARTIN R.D. (1981): Robust Methods for Time Series. In: FINDLEY D.F. (Ed.), Applied Time Series II. Academic Press. New York.

NAGEL G. and W. WOLFF (1974): Ein Verfahren zur Minimierung einer Quadratsumme nichtlinearer Functionen. Biometrishce Zeitschrift. 16, 6, 431-439.

STOCKINGER N. and R. DUTTER (1983): Robust Time Series Analysis. An Overview. Res. Rep. No. 9, Institute for Statistics. Technical University Graz.

YOHAI V. and R. MARONNA (1977): Asymptotic Behavior of Least-Squares Estimates for Autoregressive Processes with Infinite Variances. Ann. Statist., 5, 3, 554-560.

ACKNOWLEDGEMENTS:

The author wishes to thank P. DUTTER very sincerely for encouraging this work and for reading and correcting the manuscript and is grateful to K.P. PFEIFFER for many stimulating discussions.

This research was supported by the "Fonds zur Foerderung der wissenschaftlichen Forschung", project no. 4487 and no.4972.

Norbert Stockinger
Institut für Statistik
TU Graz
Hamerlinggasse 6
A-8010 Graz

Proc. of the 4th Pannonian Symp. on Math. Stat., Bad Tatzmannsdorf, Austria 1983
W. Grossmann, G. Pflug, I. Vincze, W. Wertz, eds

CORRECTION TO "UNIFORM THEORY OF COMPARISON OF LINEAR MODELS" (PROC. OF THE 3RD PANNONIAN SYMP. ON MATH. STATIST. 323-334)

C. STĘPNIAK

Lublin

Lemma 4 and Theorem 5 in this paper are in error since a step in the argument (p. 331 lines 11-13) is false. Ignoring this step we get the correct reading of the Lemma and the Theorem as follows. Lemma 4. $L([A_1,B_2],V_1) \overset{A}{\succ} L([A_2,B_2],V_2)$ iff $L(Q_1A_1,Q_1V_1Q_1) \succ L(Q_2A_2,Q_2V_2Q_2)$, where Q_i, $i=1,2$, is the orthogonal projector on $N(B_i^*)$. Theorem 5. $L([A_1,B_1]V_1) \overset{A}{\succ} L([A_2,B_2], V_2)$ iff $A_1Q_1[Q_1(V_1+A_1A_1^*)Q_1]^- Q_1A_1 - A_2Q_2[Q_2(V_2+A_2A_2^*)Q_2]^- Q_2A_2$ is n.n.d.

This correction does not affect the rest of the paper.

Department of Applied Mathematics
Agricultural University of Lublin
Akademicka 13
PL-20-934 Lublin, Poland

SUBJECT INDEX

Probability and Statistical Decision Theory

Proceedings of the 4th Pannonian Symposium on Mathematical Statistics, Bad Tatzmannsdorf, Austria, 4–10 September, 1983

Volume A

edited by

F. KONECNY
University of Agriculture, Vienna, Austria
J. MOGYORÓDI
Eötvös Loránd University, Budapest, Hungary

and

W. WERTZ
Technical University of Vienna, Austria

The Fourth Pannonian Symposium on Mathematical Statistics was held in Bad Tatzmannsdorf, Austria, 4–10 September, 1983 after the first two Symposia took place held there in 1979 and 1981, whereas the third was organised in Visegrád, Hungary in 1982. In the present volume 24 contributed papers are collected dealing mainly with purely mathematical problems. A second group of contributed papers dealing with problems of applied statistics, applied probability theory and related topics is published in a separate volume entitled "Mathematical Statistics and Applications".
The papers of the present volume concentrate on two (intersecting) topics: on the one hand, stochastic processes, while on the other hand, invariance principles: Gaussian processes, approximation of the Wiener process, distributions of spacings and of order statistics, limit theorems in triangular arrays. Furthermore, related ropics like ergodic theory, maximal inequalities, approximation of convolutions of distribution functions, stochastic programming, etc. are discussed.
The statistical contributions, as a third main topic in this volume, are even more homogeneous: amongst the submitted papers there are decision theoretic ones (treating admissibility, sufficiency, limits of experiments, etc.) and papers on nonparametric estimation (nonparametric estimators of regression curves, further rank-, L- and shrinkage estimators). One paper gives a comprehensive survey of two-sided parametric tests.
The reader will observe the close connexion between many of the questions under consideration; in particular, many results of papers which are related to probability theory have immediate applications in statistics and some belong to the common boundary of these subjects.

Audience

The book will primarily appeal to research workers in the areas of probability, statistics and statistical decision theory.

CONTENTS
to Volume A

GPSR Compliance

The European Union's (EU) General Product Safety Regulation (GPSR) is a set of rules that requires consumer products to be safe and our obligations to ensure this.

If you have any concerns about our products, you can contact us on ProductSafety@springernature.com

In case Publisher is established outside the EU, the EU authorized representative is:

Springer Nature Customer Service Center GmbH
Europaplatz 3
69115 Heidelberg, Germany

Batch number: 10607222

Printed by Printforce, the Netherlands